# Hiking

## Pure & Simple

# Hiking:

*Illustrations by the Author*

*Distributed by*
*Winchester Press*
*220 Old New Brunswick Rd.*
*Piscataway, N.J. 08854*

# Pure & Simple

by
David L. Drotar

**Stone Wall Press, Inc.**
1241 30th Street N.W.
Washington, D.C. 20007

 Printed in the United States of America.

Published June 1984

Library of Congress Card Number 83-051087
ISBN 0-913276-47-2

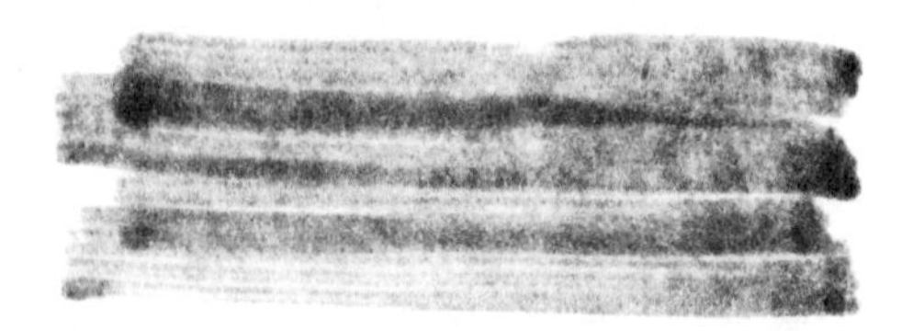

# Acknowledgments:

The author would like to thank the following people for the valuable help they provided: Pauline C. Bartel, Jackie Craven, Betty Drotar, Dorothy H. Drotar, Joshua Drotar, Karen A. Drotar, Kelly Drotar, Paul Drotar, Peter R. Drotar, Tina Drotar, Wanda Dufrene, Carol Hamblin, Joyce Hunt, Stephen P. Kistler, Kate Kunz, Peg Lewis, Arnold Madison, George Neary, Katy Remark, Donna Stillman, and Bruno Frei.

*Dedication:*

*Für Meine Schwester Karen mit Schöne Erinnerungen von der Schweiz*

# CONTENTS

# Chapter 1

# Your Own Mountain

7:00 A.M. You step outside and your senses collide with the morning. The sky is blue, the air is fresh, and a nearby waterfall chatters noisily. During the night a blanket of snow has transformed the landscape into a glistening white world.

You hoist a pack onto your back, cross a set of railroad tracks, and start to climb the gentle upward slope. Your path follows a ridge and the sun warms your body. Dripping evergreen branches frame a pair of mountain goats high on the peak above you. You reach a dry, rocky spot where the snow has melted away. Resting for a few minutes, you gaze into the valley below. A massive, craggy glacier dwarfs the village you have left. For this instant, this moment in time, the mountain belongs to you. You are hiking.

As a hiker you have the ability to see and experience parts of the world not available to anyone else, except secondhand through photographs and descriptions. A surprise awaits you around every corner. Hiking is a non-competitive activity. You don't have to worry about winning or losing because everyone who hikes is a winner.

Even though people may have different reasons for the activity, hiking is simply walking for pleasure. There are other related outdoor activities such as overnight backpacking, but ***Hiking: Pure and Simple*** was written to help you get the most from hiking itself. You may not like to camp out, but you can still enjoy hiking.

Whatever your reason for tramping in the wild—to identify plants, to photograph scenery, to get exercise—*you* set your own individual goals. You might walk only one mile behind your house or the entire two thousand miles of the Appalachian trail between Mount Katahdin, Maine and Springer Mountain, Georgia. No two trails are alike. Even two separate excursions on the same trail differ because weather and trail conditions change from day to day. Experienced hikers often have a favorite trail which they like to hike at different times of the year.

The physical benefits of aerobic exercise are well documented, and hiking fits into this category. With each step you take, the heart and lungs grow stronger and overall stamina improves. You don't have to reach the top of a mountain, however, to gain a feeling of accomplishment. You can hike all levels of terrain—not just the vertical—and thoughts of treacherous rock scaling expeditions should not intimidate you. The serene, flat path around a lake is just as suitable for hiking as a steep mountainside trail. Each will freshen your outlook and expand the capabilities of your mind and body.

# Chapter 2

# Hiking Philosophy

## *The Joys of Hiking*

Everywhere you turn these days someone is telling you why you should do something—and there's a book to tell you how to do it. Walk into any bookstore and you'll see manuals on how to cook, how to make love, or how to extend your life. In the sports and recreation section, authors tout skiing, jogging, and weightlifting. So what's the big deal about hiking? Aren't there enough camping and backpacking books already?

Well, this book won't talk about camping or backpacking. People have associated these terms with hiking for too long. Even though in many cases they are done together, they don't *have* to be. Nowhere in the book will I tell you how great it is to lie on the hard ground inside a damp tent. And if I inadvertently start espousing the benefits of rainy weather, that doesn't mean I think you should sleep in it. Fair enough?

Now that I've narrowed down what we're going to cover, I'd like to convince you why hiking is better than other activities, or, if you've never been hiking, why you should at least give it a try.

Look around. Most Americans use their free time to pursue some competitive sport like bowling, tennis, or golf. These and many other sports are fine activities, but something is wrong when an activity does not produce the desired rest or relaxation and becomes a disguised extension of the person's work. Perhaps the participant took up the sport merely because it was popular or it was something he *thought* he should do. After knocking himself out trying to be the best, is he having fun? Or is he engaging in "joyless lesisure"—a term being used today to describe the tendency to engage in too-competitive, unenjoyable pastimes.

One of the ways out of this psychological trap is to participate in non-competitive physical activities such as cross-country skiing and, you guessed it—hiking. Of course, competitive instincts can creep into just about any pursuit. But as you hike along that ridge and drink in the distant magenta peaks at sunset, who cares how many miles you've traveled or how many pounds you've carried?

This non-competitiveness leads to an effortless camaraderie among fellow hikers. For those of you who have already done some hiking, did you ever notice how easy it is to strike up a conversation with someone you meet on the trail? Liken it to anglers exchanging their latest theories about fishing techniques; there seems to be something about the outdoors that allows people to open up. I'm not saying it can't occur elsewhere, but I've never experienced this attitude in a shopping mall. Knowing that they share common non-competitive interests, perhaps hikers feel less threatened or suspicious of each other. They truly belong to a single community.

As I found out recently, the members of that community may even cross paths unexpectedly. One weekend I decided to do some hiking in the White Mountains of New Hampshire. The purpose of my trip was not companionship, but simply a change of pace; so I chose a relatively obscure trail which I felt I could comfortably handle in an afternoon.

I had hiked to the top (or the summit, if you want to use the official term), spent some time enjoying the view, and was on my way back down the trail when I met another hiker coming toward me. She gave a friendly "hi" as hikers frequently do, and we passed. Realizing that I must have just come from the summit, she called after me and asked how much farther she had to go (another thing hikers frequently do).

We faced each other again and out of the blue I said, "Do I know you?"

Now that line may still be popular in bars, but as soon as it left my mouth, I knew it didn't sound quite right here in the woods.

"No, I don't think so," she said, but the longer we stood there, the more sure I was that I knew her from somewhere. Suddenly in that flash of recognition which brings goosebumps, we both knew that we had met the previous summer in Switzerland. And now here we were again on an isolated trail far away from either of our homes.

Aside from bumping into old acquaintances now and then, one of the real joys that hiking can offer is solitude. In an age when there are so many demands on our time, we often fail to leave time for ourselves. Being alone may be difficult to achieve at your house or apartment and it is transitory at best. But there are no ringing telephones in the back-country. If you're the philosophical type—and aren't we all now and then—the concept of solitude can take on a special meaning as you hike. When you physically remove yourself from your everyday surroundings, it

KNOW ME

becomes easier to look objectively at yourself and evaluate your life. Standing on a peaceful mountain overlook, you may "see" yourself down there in the office, the school, or whatever. As you climb back into your skin at the end of the day and resume your identity, you have a fresher outlook on who you are, and are in a much better frame of mind to accept that status or make modifications.

Don't think that the benefits derived from hiking are strictly limited to personal gain. How about a family hike? What better way to bring family members together and strengthen generational ties? When the young and old hike together, nice things happen. I believe that a natural respect develops because each sees the other in a non-stereotyped way. One of the most heartwarming moments for me on the trail was meeting a couple who were engaging in a forty-sixth wedding anniversary hike. They had returned to the site of their honeymoon, but were somewhat embarrassed because the passage of time had altered their perception of the trail's difficulty. They didn't think they would be able to finish the long climb up the mountain. As a "kid" in his thirties, I applaud their spirit and initiative and hope that I might retain such enthusiasm over the years.

## *Therapeutic Benefits*

I hope the subheading for this section won't mislead you. I'm not going to claim that hiking cures cancer or stops any other deadly diseases in their tracks. But medical researchers and experts in the social sciences are beginning to present evidence that outdoor activities can have a positive effect on certain aspects of our physical and mental health.

Even the expression "go take a hike" implies a therapeutic escape from daily pressures. Solving a problem of huge dimensions or simply sorting out your thoughts can be aided by a long walk in clear air. A recent study at Purdue University has shown that exercise increases the function of the left side of the brain. Why is this significant? The left hemisphere is responsible for speech and thought. In other words, what we define as "intelligence" is centered in that half of the brain, and stepping up whatever goes on in there will turn the person into a more logical individual.

The experiment involved a group of men whose average age was forty-two. Half the group exercised three times a day for at least one hour, while the other half performed no physical activity. At the end of four months, the participants in the study were given tests designed to measure how well the left hemisphere was working. The athletic group scored higher than the non-exercisers. If we can draw any conclusions from the researchers' results, it appears that physical activity can improve the mind. The news may not be revolutionary, but I'm all in favor of something that can make you smarter and doesn't cost a cent.

Other equally interesting, but less formal findings have been receiving popular coverage. Dr. Marilyn Jean Mason of the Family Therapy Institute in St. Paul, Minnesota, has some valuable insights about the effects of outdoor pursuits on interpersonal relationships. When a marriage is on the rocks, the couple ought to climb them, she advises. During a four-day rock-climbing expedition to the north shore of Lake Superior, Dr. Mason led a group of ten husband-wife climbing partners. Because the ascents and descents over steep rock faces frequently placed the individuals in vulnerable and sometimes dangerous situations, they learned to place a high degree of trust not only in themselves, but in their spouses as well. Communication was essential. One particularly grueling climb led one couple to discover how much they really cared about each other. Prior to the experience their marriage had been headed toward divorce. Other climbers said they gained a sharper perspective of how an unwitting mistrust had prevented them from true intimacy in everyday situations.

Your own relationships may not have reached such a critical stage, but do consider ways you might capitalize on the outdoor experience to enhance some aspect of your life. Use a hike to approach a delicate subject with a friend or relative. Does your sagging self-confidence need a boost? Take a hike. Challenging outdoor adventures are being used as therapy for all kinds of people—from juvenile delinquents to business executives.

Therapeutic mechanisms may operate in unusual ways that we are only beginning to understand. While I'm writing this book, a friend of mine is traveling around the world. She just completed a trek in the Himalayan mountains of Nepal and a recent letter from her contains an anecdote about the unusual release of stress on her hike. She recalls, "We hiked twenty-two days from Dumre to Manang through the area where Tolkien got his inspiration for ***The Hobbit***. Each day we hiked four to eight hours with a twenty-five pound pack. Right now I feel really fit.

"The scenery was more than spectacular at Manang; the village faces a massive wall of the Anapurna range that seems so close that you think you could reach out and touch the mountains. Going over the Thorang pass was both an exhilarating and scary experience. Making the steady climb, we saw a rock that had 'God, give me air' written on it. Most of the trail was free from snow but a couple of areas a few hundred yards in length were ice and snow covered and at those areas the trail was about one foot wide and at the edge of a 100-200 meter drop. Some silly tune of 'one false step' kept humming through my head.

"On the day we crossed the pass, I was in the lead when we reached the first icy stretch. I took it really slowly because I was scared. When I hit the dry patch of the trail I turned to my partner, Ann, and said, 'Boy was I scared,' and then suddenly tears just started flowing. Two acquaintances had caught up to us and one had yelled at me to keep moving or something like that, but I didn't feel that had affected me. Along with the tears came a huge spurt of adrenaline because I was able to climb straight up about one hundred meters at a quick pace that just left Ann in my tracks. We were at 15,000 feet so the air was really thin."

Scientists now believe that emotional tears—as opposed to tears produced in response to irritants—may actually rid the body of harmful stress-induced chemicals. Could this hiker's combined emotional and physical status have helped her tap hidden body resources? She reports that a second crying episode occurred during the descent. "I was so happy to have made it over the pass and to have marveled at nature and been so close to the mountains that I was crying with joy on the way down."

# Chapter 3

# Body Basics

## *Training for Hiking*

A football player must be in top form before he goes onto the field. He invests many hours in a rigorous training program. Hiking, on the other hand, is an on-the-job training activity. There are no special skills required and even if you're not the athletic type, you can still be a successful hiker.

How does hiking help you achieve fitness?

Many doctors and experts in human performance no longer consider a flat assessment of height vs. weight to be a valid indicator of your proper weight because these charts do not take into consideration the percentage of body fat. A "heavy" person, for example, might harbor very little fat in his or her tissues.

The idea may sound too good to be true, but you can make a little-known principle of body metabolism work for you and lower the percentage of body fat by hiking. This is because your cells metabolize fat most efficiently during exercise performed at slow speeds. You don't need to push yourself to exhaustion in order to benefit from the effects of exercise. In addition, Dr. David L. Costill, director of the Human Performance Laboratory at Ball State University in Indiana, says that your body continues to metabolize fats even after you stop exercising. Once the process has been set in motion, it will continue to operate for two or three hours after the actual physical exercise ceases. Not bad, huh?

Although no rigid training regimen is necessary prior to attempting your first hike, you should not expect to do ten uphill miles right off the bat. Start out slowly and build progressively to longer and more strenuous hikes. A two-mile walk through a pleasant woodsy setting makes a good introduction to the art of hiking. Your next outing could be a slightly longer trek over rolling terrain. These short excursions may seem too simple for your abilities, but they will serve the purpose of giving you a good perspective of time and distance without overtaxing you physically. Nothing will turn you off to hiking faster than being stuck in the wilderness—cold, tired, and hungry—because you've wandered too far from home.

Should you undertake a jogging or running program? Let's face it. Except for vacation periods, most people are not able to spend day after day hiking in the wilderness. But running *is* something they can easily integrate into their normal routine at home. Many hearty backpackers use running as a supplementary sport to strengthen heart, lungs, and muscles of the legs and back. They are able to hike faster, further, and bear heavier packs with less difficulty.

However, even author James F. Fixx of the best-selling ***Complete Book of Running*** admits that running does have several limitations. As an overall conditioner it falls short because it neglects flexibility and does not develop muscles. If you do too much running, claims Fixx, you can actually diminish your hiking ability because the stress leaves your body fatigued. A balanced fitness program for those who want to maintain optimum efficiency between hikes might consist of a limited running schedule combined with other exercises.

As in any form of exercise, build your running program slowly. If you've never run before, a quarter-mile is plenty on your first day; no more than one mile if you're already in fair shape. Break up this distance into several intervals of running and walking. Then, as your endurance increases, do less walking and more running until you are eventually able to run the entire distance without stopping. Your ultimate goal should be about three miles.

There is no point in running faster than you are comfortable in doing because you will only get tired more quickly and stop sooner. Remember too, faster is not any better than a moderate exercise pace. It is not the speed, but the total distance you cover which determines the conditioning effect of the activity. A good rule of thumb concerning pace is this: *If you are not able to carry on a conversation without shortness of breath, you are going too fast.*

Finally, ease up on your program several days in advance of your planned hike. Run only one mile two days before the big day, and do no running the day before your hike.

For those who live near one of those fancy health clubs and can afford the membership fee, such facilities provide an excellent way of getting in shape. The *Nautilus* machines work on the concept of exercising muscles over their entire range of motion to increase flexibility as well as strength. There are special exercises for each part of your body, and when you do the prescribed workout properly, you also gain a cardiovascular benefit.

However, you can duplicate many of the same exercises at home without special equipment. The next section presents several simple exercises which will increase flexibility and help strengthen those areas of the body used while hiking. Whatever pre-hiking regimen you opt for, a properly conditioned body and the excellent conditioning effects of hiking itself will enable you to gain the maximum enjoyment from your wilderness hiking experience.

## *Exercises Before Hitting the Trail*

Here's a set of starter exercises you can do in the comfort of your own home or yard. Although there are no special limitations for age or sex, overweight persons may find some of the movements more difficult than others. Don't get discouraged. Do as many repetitions as you can, with a goal of 10-15 in mind. Form is more important than quantity.

### *Calf Stretch*

Hikers may experience a tightness in the calf due to the muscle's repeated use over a limited range of motion. To counteract this effect, a simple precaution can be taken. The muscle must be stretched to offset any tightening forces.

Perform the following exercise before and after your hiking sessions. Stand on a set of stairs with both heels hanging off the edge of one step. Keep your balance by placing your hands on the railings or against the wall. Now, slowly lower the left heel below the level of the step. As you bend the right knee forward, you will feel the resistance offered by the calf muscle. Lower the heel as far as you can and keep your weight on that leg. Gently rock forward and backward for twenty seconds. Do not bob up and down or perform the exercise in short, quick movements as that will tend to produce the opposite effect—tightening instead of stretching. Now do the same exercise with the other leg. Repeat the set several times.

If you aren't near a step, don't despair. Simply place one leg in front of the other and shift your weight onto the forward leg. Reverse legs and repeat.

### *Hamstring Stretch*

You should also stretch your hamstrings (the group of tendons and muscles behind your knees and upper legs) since these are also tightened during the hiking process.

1) Stand relaxed with your feet flat on the floor and shoulder width apart. Bend forward from the waist, letting your arms hang loosely. Keep your chin tucked in and allow your neck and arms to relax. Breathe slowly and deeply through the nose. Trying to keep your legs as straight as possible, touch your toes with your fingertips. Hold this position for about thirty seconds or longer if possible. Now stand up again and repeat several times.

2) Sit on the floor with one leg straight forward and the other at a comfortable angle. Bend forward from the waist and grab the toes of the outstretched leg, keeping the knee pressed as close to the floor as possible. You may find it difficult to not bend the knee. Hold for twenty seconds and repeat with the opposite leg.

## *Lower Back Stretches*

Don't ignore the all important back because it may receive considerable stress if you're carrying a pack on inclined surfaces. Correct posture and a strong, limber back will enable you to handle the extra strain and prevent serious problems from occurring.

1) Lying flat on your back, point your toes straight. Then bend your left leg. Lock your hands around the leg just below the knee and draw

the knee to your chest. Squeezing tightly, hold this position for several seconds. Release the tension and squeeze again. Repeat several times with each leg.

2) Lie on your back with arms stretched above your head. Lift your legs and hips and slowly lower them back over your head. Try to keep your legs as straight as possible. Hold for one minute.

## *Torso Twists*

1) With your arms outstretched parallel to the floor, hold a broomstick behind your shoulders. Spread your feet shoulder width apart and stand straight. Now rotate your upper body as far to the left as possible while keeping the legs and hips locked in the forward position. Turn to the right and alternate direction back and forth.

2) Hold a broomstick as in the first exercise, but bend forward at the waist. Keeping a firm stance, rotate your upper body from left side to right side and vice versa, similar to the previous exercise. Your left arm will move down as your right comes up.

*Bent-over torso twist*

The following group of exercises is designed to strengthen specific areas of the body:

### *Bench Leg Raises*

Exercises which strengthen the thigh muscles (quadriceps) are useful to hikers because these muscles propel you up hills. However, many of the exercises commonly performed create stress on the knee joint. Here is a good substitute which works the quadriceps without invoking the repeated flexing of the knee.

Sit on a wide bench (such as a picnic bench) with legs stretched in front of you about eighteen inches apart. Grasp the rear edge of the bench, palms facing forward, and lean back slightly. Now raise both legs as high as you can while bringing your feet together. Hold this position for a few seconds and then slowly lower your legs to their original position.

### *Broom Bend Overs*

This exercise works the lower back. Stand erect with feet shoulder width apart and a broomstick held behind your shoulders. Keeping your legs straight and your head up, bend forward at the waist as far as possible. Hold for a few seconds and then raise yourself to the upright position.

### *Shoulder Shrugs*

You'll need a pair of dumbbell weights for this exercise. Stand straight, holding the weights at your side. Slowly raise your shoulders as high as you can, then pull them back, and finally lower them to their starting position. The action should be performed in one continuous circular motion over several repetitions.

### *Pelvic Tilt*

Stand flat against a wall with your head, shoulders, butt, and heels touching the surface. Without allowing these body parts to break contact, pull in your stomach and press your lower back to the wall. Practice this routine several times a day. When you have mastered the technique you will be able to perform the pelvic tilt without a wall to line up your body parts. This exercise promotes the most efficient body posture by toning the pelvic muscles. Notice how the action relieves tension on the lower back.

### *Alternate Leg Raise Situps*

"Situps? Wait a minute," you groan. "Of what benefit can *they* possibly be to hiking?"

No, I haven't suddenly developed a sadistic streak. As mentioned previously, a healthy back and good posture are important to hikers, especially when bearing heavy packs. The stomach muscles help to keep the spinal column aligned properly. An out-of-tone and flabby stomach can contribute to lower back pain in two ways. One, the muscles are not doing

*Alternate leg raise situps*

their job in supporting the spine; two, the excess weight pulls the spine in the wrong direction.

Perform these alternate leg raise situps every day as part of your regular exercise routine. Lie flat on your back with hands grasped behind your head. As you sit up, raise your left leg off the floor, bringing the knee and your right elbow together. Lower yourself slowly to the prone position and repeat the exercise with the right leg and left elbow. Do as many repetitions as possible.

## *Bent Knee Situps*

This version of the familiar somebody-hold-your-ankles situp looks like you're cheating, but it is actually more difficult and will give the stomach muscles a much harder workout. Lie on your back with legs bent at the knee, feet flat on the floor, and hands locked behind the head. Now sit up and touch one elbow to the opposite knee. See what I mean?

## *Push-ups*

With the strong emphasis on toning muscles in the legs and back, it is easy to neglect the upper body. Although you might think that the arms, shoulders, and chest are not used while hiking, this is not entirely true.

Drooping shoulders make it more difficult to maintain proper posture and balance. Carrying a pack becomes more difficult.

No exercise could be simpler than the basic push-up. A faithful routine will do wonders to keep the arms, shoulders, and chest in good shape. Lie on your stomach with your hands placed palm downward beneath your shoulders. Keeping your body as straight as possible, push up. Women are allowed to use their knees as a fulcrum and bend from that point. Slowly lower yourself to the ground and repeat.

# Chapter 4

# Pack It In

## *The Pack*

You see them in the woods, on the beach, and on the street. Everyone seems to own a pack these days and perhaps you're thinking about buying one. A pack is an efficient way to carry your belongings. There is also an extra benefit. Your hands remain free to take a picture, scribble a note, or push a branch out of your way.

How do manufacturers define the different types of packs? There are three main categories. The class depends on the size of the load which the pack holds.

Many people glance at a cloth case on someone's back and immediately call it a "backpack." Avoid confusion by using this term only when speaking of a large pack with a rigid support system (an outer metal frame or internal stays).

Day packs are small packs which carry items for a day's hike, but will probably not hold enough gear for an overnight stay somewhere. They have many in-town uses as well, such as carrying books, store purchases, or your lunch. A good day pack will cost about $25.

Rucksacks are larger than day packs but not as big as full backpacks. They do not have rigid support systems. On group outings, a parent might carry a rucksack with food and clothing for other family members. Or you

*A small day pack will hold items for short hikes.*

might use a rucksack on a winter hiking excursion when you need to carry heavier clothing. Expect to pay $40 and up for a rucksack.

Pack style and capacity varies greatly, but the important point to remember is that you should select a pack according to your needs. Don't allow a huge collection of buckles, zippers, flaps, and pockets to dazzle you. Do you really need those six pockets when you only plan on carrying a bottle of water to a trail's summit?

If you do decide on a full backpack for your longer trips, there are several construction features to consider. You can choose either an external frame or an internal frame. External frame packs offer the most roominess. They have upper and lower bars which provide additional space for tying items to the outside. The bulky external frame packs, however, are somewhat inconvenient for train travel or loading in and out of a car. In rugged country, they're "about as graceful as wearing a lawn chair," says experienced outdoorsman Rob Schultheis in *Backpacker* magazine.

If you're willing to sacrifice some space, the internal frame pack offers a good alternative. The manufacturer has sewn a pair of flexible stays into the inside of the pack for support. The pack hugs your back and bends as you move. Most internal frame packs do have straps for lashing extra cargo to the outside.

Although they are called **back**packs, this does not mean that the entire

*This backpack has an internal frame. The wide, padded hip belt cushions your hips from heavy loads.*

load rests on your back. The pack distributes the weight over your shoulders, back and hips, with the greatest portion on the hips. So choose a backpack with a wide, comfortably padded hip belt. Reject any backpack with shoulder straps as the only means of suspension.

A main compartment which opens from the front, rather than the top, is a popular feature in the newer designs. You load the pack like a suitcase. When you want to remove a single item, you don't have to dig through everything else.

Prices for backpacks start at $70. The same backpack model may come in different sizes. If you have an average build, get a size Medium pack. If you are smaller than average, buy a size Small, and so on. Try on the pack in the store and ask the salesperson to put in some weight. Make sure that you can easily make all the adjustments yourself.

No pack can make forty pounds weigh less. But a well constructed, proper fitting backpack can help you enjoy many miles of travel.

## *What to Carry on the Trail*

The great freedom and mobility that hiking offers can be lost if you are overburdened on the trail. Although a pack provides an efficient method of carrying many supplies, the wise packer will travel light and try to stick with the essentials. This doesn't mean that you should necessarily exclude items which aren't crucial to your survival or comfort. It never hurts to bring along your fishing rod if you think you might get the opportunity to use it.

Clothing and food will take up most of the space in your pack and this book contains chapters dealing with each of these topics separately. Here's a roundup of other items which you may want to include on your trip.

### *Sun Protection*

Although you should protect yourself adequately against the sun anywhere you hike, your skin is particularly vulnerable on mountainous treks. The thinner atmosphere filters fewer of the sun's intense rays before they reach you. Snow fields at these higher elevations compound the effect by reflecting the rays instead of absorbing them. The cooler air is deceptive because you don't realize you're getting burned.

Use a good sunscreen lotion—not merely a suntan oil—over the exposed parts of your body. Lip emollients now come with similar sun-blocking compounds which will prevent painful blistering to this often unprotected facial area. Don't forget to protect your eyes. I always slip a pair of sunglasses into my pack, whether or not the day is sunny. You know how disorienting road glare can be when you're driving. The same effect can take place on a hike under open skies. Neutral lens shades of gray or green are best; they're the least tiring on the eyes.

## *First Aid Kit*

Chances are you'll never use a first aid kit, but it's a nice feeling to have one tucked away in the corner of your pack to take care of minor injuries such as cuts, scrapes, burns and blisters. You can buy first aid supplies in a drugstore and assemble your own kit, or buy one of the packaged kits now being marketed with the outdoorsman in mind. The latter is really not much more expensive than the sum of the individually purchased items, plus you get a case to hold everything neatly. These kits contain an assortment of dressings, adhesive strips, tape, antiseptic, and moleskin. If you have a medical problem (allergy, diabetes, etc.) for which you have prescription drugs, you can and should include them in the kit. Also, consider adding a snake bite kit in your pack if your hike takes you through territory populated with poisonous snakes. Become familiar with its use *before* you need it because every second counts when you are bitten.

## *Whistle*

A simple whistle takes up practically no room in a backpack, but could save your life. If you ever become critically injured or lost, use the whistle to summon emergency help. You could wear one around your neck to save fumbling around. This is a good idea for kids—to be used *only* in an emergency.

## *Pocket Knife*

I'm rather prejudiced when it comes to recommending a pocket knife. You see, I spent several months in Switzerland and I never went anywhere without my trusty Swiss Army Knife. The knife is great for hacking off a hunk of cheese, bread or fruit, or prodding a stubborn juice carton open. Made of high grade stainless steel, the blades glide into position at the flick of your fingernail.

The knife is extremely compact and lightweight, but the feature that makes the Swiss Army Knife so famous is its versatility of functions. The gadget is really a miniature tool box containing screwdrivers, can opener, bottle opener, wire stripper, leather punch, corkscrew, and scissors, among other things. Some models have a magnifying glass that you could use to start a fire if the sun is out.

## *Ziplock Bags*

If I had to choose the ten best inventions of the last century, ziplock bags would rank right up there with silicon chips. I always keep an ample supply of bags on hand to encase just about everything before it goes into my backpack. Not only do the watertight plastic bags keep items dry, but they help organize the contents of your pack by forming logical groupings. Notebook and pen go in one bag. For overnight excursions, toothbrush and toothpaste in another. You can carry paperback books, matches,

*A Swiss Army Knife is a useful companion on any hike.*

*kleenex,* film, and many other personal items without any fear of them getting wet or leaking. Ziplock bags are also ideal for individual portions of food.

## *Binoculars*

These fall into the category of "trail frills." While not necessary for you to conduct the journey, binoculars can increase your enjoyment immensely by extending your vision. I'm not talking about the old-fashioned clunkers which wrenched your neck everytime you wore them to a football game, but rather the new ultralight versions with precision lenses. Once you transform a white speck on a distant mountainside into a real live sheep or watch bald eagles soaring in the treetops, I guarantee you'll be hooked.

# Chapter 5

# The Feet: Base of Operations

## *Choosing a Hiking Boot*

Boots are the most important piece of equipment for backcountry use by far. Properly shod feet make the difference between a good time and misery. This does not mean that running shoes have no place on the trail, they are relatively inexpensive, very convenient for short hikes, comfortable, and do not tear up the trail as lug soles do.

Although running shoes and other kinds of sneakers are perfect for smooth, dry ground, they are not practical on more primitive trails. They get wet easily and offer no protection against sharp rocks and rough sticks. Street shoes encase the foot with sufficiently sturdy material, but their soles are much too slippery to provide firm footing on unfamiliar terrain. You need something specifically designed for the hiker.

So you've decided that you want to buy a good pair of hiking boots for the trail, but now the difficult question arises. *Which* pair should you buy?

The first step is to go to a reputable backpacking shop with a knowledgeable staff. Bill Kemsley, publisher of *Backpacker* magazine, warns that "there are lots of shoes in shoe stores that *look* like hiking boots, but they

*Choose footwear appropriate to the trail surface.* Docksiders *are a comfortable choice for wooden boardwalks.*

can let you down when you're on the trail." This is not the time to economize on quality. You can expect to pay over $50 for boots.

Every hiker would like a boot that is 100% waterproof, but the only truly waterproof boot material is rubber. Since rubber does not breathe, perspiration cannot escape and you will end up with cold, soggy feet if you hike in all-rubber boots. Leave them for stand hunting or slow, marshy treks. The newly-developed *Gore-Tex*™ fabric *is* waterproof and breathable, but boots made from it contain leather sections and numerous exposed seams where the leather and *Gore-Tex* are stitched together. These seams are potential points of leakage and must be treated with a waterproofing compound just as other leather boots. Even though no leather can be made totally waterproof, most hikers still prefer an all-leather, non-suede boot with as few seams as possible. The seams should be double or triple stitched for strength.

Examine the quality of the leather itself. The texture should be heavy without being stiff and hard. If the boot has the smooth kind of finish, roll the leather tightly between your fingers. Reject the boot if you see cracks in the finish. The "upper" is the entire part of the boot which lies above the sole. Highly recommended is a fully leather-lined upper; that is, two thicknesses of leather in the construction. The soft inner lining pampers your feet, while the double layer insulates, gives better ankle support, and wears longer than single wall construction.

The insole (the part on which your foot rests) should also be good quality leather. If you see pressed leather fibers or other material, be wary that the manufacturer may be trying to cut corners.

The best materials are wasted if the leather is weakly attached to the sole. The term "welting" is used to specify the method of attachment. Ask the salesperson which type of welt is used. Norwegian and Littleway welts are the strongest types and employ rows of stitching to fasten the uppers to the sole. Other methods, including cementing and injection molding, may come undone and shorten the life of your boots.

Take care when selecting the type of ankle closure. Padded areas called "scree collars" were invented to provide comfort and keep stones, gravel, water, and snow from slipping inside the boot. However, the design often failed to serve these purposes and the material—most likely vinyl—became worn and irritating to the leg. Insist on a leather scree collar if it is exposed. A better design is now available in a few boots. The padding has been moved entirely to the inside, while the exterior forms one continuous surface.

You can tell a lot about a boot by the placement of the tongue area. If it is close to the front of the boot, the manufacturer may be trying to skimp on leather. This can create several problems: squeezed toes when the laces are drawn, large wrinkles exerting pressure on the toes, and exposed seams vulnerable to rocks and sticks. Make sure the tongue opening is placed well behind the ball of the foot.

Don't "overboot" yourself. Heavier boots may not be necessary for the kind of hiking you'll be doing. Large, heavy lug soles are often not necessary for most trail hiking. You might do well with a pair of 6-inch high lightweight composition sole boots, or with a pair of medium weight trail boots. I've found that a pair of well-worn work boots are good for hikes over dry terrain. They're cheap and will last forever.

It only makes sense to try on the boots with the socks that you will wear while hiking. Two pairs of socks cushion your feet, wick away moisture, and prevent blisters. Wear a snug, lightweight pair next to your skin, and a heavier wool pair over these. Any sliding will occur between the pairs of socks, rather than between the materials and your skin.

Once the boots are on your feet, ask yourself one question: Are they comfortable? With the laces firmly tied, walk around for a few minutes. You should not feel any pressure points on your feet. You should be able to move your toes up and down freely. Do deep knee bends and squat thrusts to be sure your heels remain firmly in place without slipping. (And make sure the store will exchange the boots if they don't fit.)

During a ten mile hike, a 125-pound hiker will strike over two million pounds against the bottoms of his feet. Comfortable and durable boots will absorb such stress and carry you over many miles of hiking enjoyment.

*Sneakers offer no ankle support on rocky trails.*

*Note how a sturdy pair of hiking boots can prevent the ankle from turning.*

# *Caring for your Hiking Boots*

The purchase of good quality hiking boots represents a considerable investment, and proper care will prolong the life of your footwear. Before the soles ever meet the trail, you should begin a program of maintenance.

As soon as you bring the boots home, waterproof and condition the leather with a compound recommended by the manufacturer. The type of tanning process used on the leather will determine which dressing you should use. The general rule is to use mink oil on oil-tanned leather, and a wax base compound such as *Sno-Seal*™ on chrome-tanned leather.

Rub the dressing into the leather with your fingers and allow it to soak in. If the weather is sunny and warm, do this outside and the warmth of the sunlight will help the leather absorb the compound. Repeat the process until the boots absorb as much dressing as possible; then wipe off the excess with a cloth.

Do not leave excess amounts of waterproofing compound on the surface. You might reason that the extra compound would be a good barrier against moisture, but in fact it will collect dirt. The dirt will absorb the water which you are trying to keep away from your feet.

The welt (the area where the boot is attached to the sole) is particularly vulnerable to moisture because of the heavy thread and large stitches. Waterproof this area thoroughly, but do not leave any excess compound. You might want to buy a special welt sealer. It fills in the tiny openings and makes the welt watertight.

The next step may surprise you. After you have waterproofed the exterior of your boot, don't wipe your hand on a rag. Use the inside of your boot to clean your hands of the compound. It is just as important to treat the inside leather because it gets wet also.

You will be anxious to put your new hiking boots through the rigors of a long hike, but a little patience now will pay off later. A gradual break-in period will ensure that the heavy, awkward feeling will give way to maximum comfort. Wear the boots for brief periods of time (around the house and on short walks). Increase the distance slowly. Francis Limmer, vice president of the New Hampshire bootmaking firm Peter Limmer & Sons, Inc., suggests that the new boot owner stay away from the high peaks and make three or four easy hikes over rolling terrain. The slow break-in process allows flexible bending points to form as well as permitting the leather to conform to your foot.

You wouldn't let mud and sand dry on your skin overnight, and letting this happen to your boots is just as careless. Dry debris pulls oils from the leather and hastens the aging process. The leather soon breaks down and cracks. To prevent this from occurring, clean dirty boots with a soft brush and water. Saddle soap can be used if necessary. Let the boots dry slowly in an airy spot—never in the oven or next to the fireplace. When they are dry once again, recondition the leather with the compound you used on the new boots.

Don't overlook the small details. Examine your boot laces periodically. If they are worn or frayed, replace them with a new set. You will avoid a broken lace while you're in the wild, as well as making your boots look better. Are your socks clean and dry? Hiking with grimy socks contributes to the salt and acid buildup on the boot's interior. This will make leather fibers and stitching rot more quickly. Always start your hike with clean socks and carry a fresh pair with you. Your feet and boots will both benefit.

## *Your Feet on the Trail*

Unlike a pair of boots which requires attention between uses, your feet need constant care both on and off the trail. If you follow a few precautions, you can substantially reduce your chances of discomfort or injury.

Keep your toenails clipped as short as possible. Nails which have grown beyond the tips of the toes will bump against hard boot leather with each step and tug at the softer underlying tissues. The slight impact may be imperceptible, but repeat this torturous action over several miles and you'll be screaming for mercy by the end of the day.

Dirty socks can harm not only your boots, but your feet as well. As the material mats down, collects dust and becomes stiff, it rubs unevenly against your skin creating sore spots and blisters. The same action happens when a stone or twig gets into your boot, but the threat is more immediate because your tender, perspiring skin is susceptible to sharp objects. Remove the foreign body immediately, before it has a chance to do any damage.

The soles of your feet contain a greater number of sweat glands than any other part of your body except your hands. During hot weather a refreshing foot bath in a pond or stream will cleanse the skin and rejuvenate aching muscles. Be sure to do this when you have at least half an hour, such as when you stop for lunch. Feet swell when they're first taken out of the boots and it takes a while for them to shrink down again. If you aren't able to spend this much time or there is no source of water, you can do the next best thing by lying on your back with your boots elevated for a few minutes.

Despite the most carefully placed footsteps and the sturdiest of boots, your ankle could suddenly twist on an unstable rock and cause a painful sprain. If this happens, know what to do. "Failure to properly treat a 'simple sprain' can lead to serious problems later," says Dr. Robert Athanasiou, the assistant director of emergency services at Samaritan Hospital in Troy, New York. "Only twenty to sixty percent of patients are free of symptoms one to four years after injury. The average disability is from four to twenty-six weeks, and there is a very high incidence of recurrent sprains."

Specifically, a sprain refers to an injury to a ligament, the connective tissue between bones. (Don't confuse this with a *strain*, which involves muscle.) If the outside ankle swells and is tender to the touch, you've injured the most commonly sprained ligament in the body. Such a sprain can range in severity from first-degree to third-degree.

Although any sprain can be painful, first-degree sprains are the least serious because the mishap has only stretched the ligament and not torn any fibers. Usually you can bend the foot normally. Second-degree sprains involve partially torn ligaments, resulting in more pain. Walking is difficult and some bruising may occur due to ruptured blood vessels. The most severe type of sprain is third-degree in which ligaments are completely torn. Within two or three hours after the accident happens, ankle swelling increases to the size and shape of an egg.

Treatment of a sprained ankle should begin immediately. Place some ice or snow, if available, inside a plastic bag and hold the cold compress against the injury. As an alternative, use water from a cold lake or stream. Keep the ice pack held against the ankle for at least one hour, but never allow the area to become painfully chilled. The last thing you need is frostbite. Regulate the temperature by slipping a bandana or thin sock between your ankle and the ice.

The next step is compression of the joint to reduce swelling and limit the range of motion. Wrap a bandage or any clean cloth cut into wide strips around your entire foot beginning at the toes and working your way over the ankle. Wrapping only the ankle reduces circulation to the rest of the foot. A healthy pink color after pinching a toe between your fingers will assure you that circulation is still okay.

Slip your boot over the bandaged foot and lace it snugly but not tightly. You will now be able to walk to the trailhead if you move slowly and use a walking stick to keep the weight off the injured foot. Once every hour, stop for a ten-minute rest period during which you should re-ice the foot and elevate it above your hip to counteract blood pooling in the lower limbs. Proper blood flow will reduce swelling and hasten the healing process.

Continue the ice, compression, and elevation treatment when you get home and take aspirin for pain relief. For serious sprains, consult your doctor who will be able to prescribe stronger pain killers and anti-inflammation drugs, as well as determine if other treatment is necessary.

# Chapter 6

# What to Wear

## *The Trail Wardrobe*

One of the nicer aspects about hiking is that you don't need to spend a lot of money outfitting yourself with brand new clothes. The only dress code that nature enforces is practicality. In fact, you probably have enough items in your dresser right now to go out hiking; so this section will be more of a guide on how to put things together, rather than what to buy.

As you do more hiking, of course, you'll discover that certain clothing features are better suited to the sport and you'll gain a sharper idea of what you personally like and dislike. When you reach that point, then it's time to consider buying items specially designed for the hiker. I think you'll find that well-constructed outdoor clothing will nicely fill the bill for off-trail occasions as well. Three years ago I bought a Woolrich brand parka which I intended to use for hiking. The item was not cheap by any means, but has since served me on hikes, while traveling, and for general everyday use. I've had the jacket cleaned several times and it still looks as good as new. Although the initial price of such pieces may be somewhat higher than discount store clothing, the quality makes the more expensive piece the better buy because you get much more use out of it.

Please don't immediately run out and buy something because you see it recommended in this book, or any other book for that matter. Wait until you would ordinarily purchase a particular item to fill a gap in your wardrobe, and then consult the suggestions in the book.

One more point before we move on to discussing specific items of clothing. You're probably wondering where you buy these super clothes. The most logical place to check first is a local store which specializes in backpacking equipment. The staff generally has a lot of first-hand knowledge about their line because they are usually hikers themselves. You can try on the clothes before buying, and if you later encounter any problems with the merchandise you can take it back to the store. On the negative side, the store's selection and inventory may be limited.

Although the market for quality outdoor clothing is expanding, your area may not have such a store. In that case, check out the regular sporting goods stores. Larger establishments may even have a separate camping section, which means you're on the right track. Walk past the tents and sleeping bags and you'll be sure to find some good outdoor clothing.

Mail order catalogs are a popular alternative to on-site shopping. They offer a large selection of clothing, accessories, and equipment, and sometimes discounts are available. I've found the quality of the merchandise to be excellent. When I'm unable to find what I want in a retail store, I never hesitate to order it from a catalog. The disadvantages of this system are that you cannot try on the items or ask a sales clerk for advice. You must choose the size carefully, and if you're like me, it's a hassle returning goods through the mail and waiting for an adjustment.

Where do you get catalogs?

If you've ever subscribed to one of the outdoor magazines, your living room is probably buried with the colorful advertisements of L.L. Bean and others. Once your name gets on a mailing list, you begin to believe in those eighteenth century theories of spontaneous reproduction that your high school biology teacher pooh-poohed. Now's the time to gather up all the brochures, toss away the outdated ones, and construct a nice little reference library of manufacturers. For those unfortunate souls who are immune to junk mail or find gaps in their collection, there's an appendix of companies at the back of this book. Just write to them and I'm sure they'd be happy to send you their latest catalogs.

When assembling your outfit for a hike, there is one basic principle that will guide you in spring, summer, fall, and winter. Layering is the universal theory to which all hikers adhere, and any advice you read about clothing relates to this idea in one way or another. You've probably heard the term used before, but what exactly is layering and why does everybody make such a big fuss over it?

Essentially, the layering technique of dressing means wearing several lightweight layers of clothing instead of one heavy garment. This action helps to maintain as constant a body temperature as possible. During periods of cold weather and low activity levels, the layers trap air between each other as well as within the fabric itself. The trapped air insulates the body to keep it warm. As the body temperature increases with more strenuous activity or rising outdoor temperature, it is a simple matter to remove a layer of clothing. From the standpoint of efficiency, it is much easier to regulate body heat than to reheat the body once it has cooled off.

Build your own layered system by working your way inside out. One characteristic that your underwear should possess is breathability; that is, it should permit perspiration to pass through. This is especially important during cold weather hiking, or situations where the weather could suddenly turn colder.

One hundred percent cotton underwear is not really suitable for hiking. Once it becomes wet, cotton feels clammy and no longer insulates. Wool works well in this respect, but who can stand even the thought of those itchy fibers rubbing against bare skin? Some manufacturers have tried to overcome this objection by creating a two-layer fabric with the wool on the outside, thus eliminating direct contact with the skin. Wool lovers also have another option with blended fabrics which may be easier on the skin than the straight stuff. The new synthetics such as *olefin* and *polypropylene* are light, non-itchy, and breathable. Underwear made from these fabrics is becoming increasingly widespread and is a good choice for hikers. And lastly, old-fashioned silk fulfills these requirements and is still a favorite—but its price isn't.

These recommendations especially apply to the "long John" type of underwear. Obviously you won't wear a full suit of long underwear on a warm day, but do consider the above principles at any time of the year and adjust your wardrobe accordingly. For example, an inexperienced hiker might start out on a sunny spring day wearing a sweatshirt. As he hikes uphill, the sweat pours out of his body and soaks the shirt. At the summit, a stiff breeze picks up and chills not only the hiker's body, but his spirits as well. Worrying about getting into dry clothing, he can not fully enjoy the rest of his jaunt.

The enlightened hiker, however, has decided on a simple two-piece system for the upper torso. He wears a light, breathable shirt next to his skin and a lightweight wool shirt over that. During brisk hiking, the removal of the wool shirt prevents his sweat glands from working overtime and he remains comfortable. When the day turns cooler and he heads back downhill, he slips the wool shirt over his dry undershirt for a comfortable descent and pleasant memories of his hike.

There are really no special requirements for the bottoms of your underwear. Remember that your legs will be in constant motion for the duration of your hike. Therefore choose underpants that have proven themselves comfortable in everyday wear and definitely not something which will chafe or rub. This isn't the time to break open that new package of designer bikini briefs, no matter what your *après* hike intentions. Wait until you've worn and washed your underwear several times before taking it on the trail. The same rule holds true for most other clothing items.

In terms of outerwear, should you wear shorts or pants? Shorts offer greater freedom of movement and I like to wear them whenever the weather permits. Even when the weather doesn't permit, I'm enough of an optimist to take them along in my pack for a quick change. Perform this test to help you decide if your shorts will serve you well: Stand in front of a chair and step up. Then turn around and step down. If you have any difficulty, you know immediately that your shorts are too tight. Choose a looser-fitting style.

Besides weather and temperature, before starting out on your hike, consider whether you'll be heading through an area with undergrowth or briars. Also, don't neglect the mosquito or stinging-fly factor. For those occasions when you need to cover the entire leg, you have several pant options. Everyone has a favorite pair of blue jeans. Their comfort and durability make jeans a practical choice for certain hiking situations, but you should be aware that not all conditions favor them. In cold weather, wet jeans are next to impossible to dry and you'll feel like you're wrapped in a dishrag. A simple hike over damp grass or snow may leave you soaked because the fabric absorbs moisture like a wick. A better choice would be pants made from a cotton/synthetic blend because they shed snow and dry quickly. When the thermometer really plummets, nothing can beat wool.

There is an alternative that falls somewhere between shorts and long pants. Cross-country skiers have discovered the advantages of knickers and I'd like to see more hikers in this country wear them, too. (They're standard garb for hikers in the Swiss Alps.) While keeping most of the leg covered, knickers eliminate the problem of dragging cuffs. Without all that excess fabric below the knee, you are better able to move gingerly over rock-strewn and brushy terrain.

We've now touched on the basic pieces of clothing. You have a virtually unlimited selection of items for additional layers. Pile on an extra shirt and wool sweater, followed by a down or synthetic vest. The last layer should be a wind-proof shell. Although any tight weave fabric is acceptable, everybody's excited about the recently developed fabrics such as *Gore-Tex*™ and *Entrant*™. They allow perspiration to evaporate through the tiny pores of the material, but block the penetration of rain and snow. The end result is that you stay dry inside, and all the water ends up outside, where it belongs. With this in mind, avoid non-breathable clothing that is made from rubber or plastic. These materials are better than nothing on a rainy day, but they prevent body moisture from escaping and encourage the condensation of trapped sweat on the inside.

When you shop for a good quality blouse, business suit, or whatever, you can easily spot the difference between superior and shoddy workmanship. The same guidelines apply to outdoor clothing. A keen eye is all you need to examine construction features critically. Look for sturdy fabrics and tight, well-stitched seams. Zippers should glide smoothly. After all, if you have trouble with a zipper in the store, you can't expect that it will be any easier to operate while you're running for cover during a sudden storm. Plastic zippers tend to be more trouble-free than metal versions because they are not subject to rust and corrosion caused by moisture in the atmosphere.

Pockets are the order of the day. As far as I'm concerned, the more the better. An abundance of pockets on your pants, shirts, and jacket will help you keep items separated and readily available, preventing you from having to dig through a loaded pack to locate a map, for example. However, there is another school of thought on this matter. Extra stitching creates more holes for rain to seep through, and pockets and pull tabs can more easily get snagged on branches.

Another feature I particularly like on outdoor clothing is the *Velcro*™ closure which manufacturers are now using on cuffs, hoods, waistbands, etc. This type of closure allows you to instantly adjust the item to your own size and seal out the weather at that point. Ingenious product developers have adopted *Velcro* for a variety of other uses ranging from pocket flaps to camera bags. My hiking boots even have an adjustable *Velcro* tongue closure.

This book can't possibly give detailed reviews of all the brands of clothing and equipment on the market. The business is such a rapidly changing and competitive field that by the time the information reached you, it would be out of date. The scope of this chapter, then, is to give the

Velcro *closure allows you to adjust garment snugly.*

would-be hiker enough background to make intelligent choices about his or her clothing. If you decide you want specific product evaluations of a more technical nature, I recommend *Backpacker* magazine. Their staff tests equipment and prints excellent reviews, from time to time, on a wide range of outdoor items.

## Trail Accessories

Browsing through the sports catalogs, you will see a wide range of accessories for sale and your first reaction might be: "Gee, wouldn't that be neat. I think I'll get one." Pretty soon your order form is filled up with unnecessary items and you don't have any funds to buy that sweater you really wanted. What's worse, after a year you're left with a collection of items you've hardly used.

Don't let this happen to you. Before you buy something, ask yourself if it is really necessary. Use the following guide to help you evaluate your own needs and separate the truly useful accessory from the frivolous.

### Hats

This item of clothing is not merely an accessory, but at times a necessity. Your neck and head radiate more heat than any other body part, yet are the least protected from the weather. If your body begins to lose too much heat, an automatic safety mechanism protects vital organs by restricting blood circulation to the feet and hands. Every experienced hiker knows the old saying "if your feet get cold, put on a hat."

A simple wool ski cap is convenient and works well for most situations. On windy or rainy days, an attached parka hood protects your neck and head from the intruding elements. For added warmth wear the ski cap underneath the hood. Severely cold days might require even more protection in the form of a face mask hat, scarf or balaclava. The latter is a one-piece combined hat and scarf that looks like it came from the Arabian desert.

Besides keeping warm, there is another good reason for wearing a hat, namely keeping cool. Use a lightweight, light-colored hat to prevent the intense rays of the summer sun from baking your noggin. If you're hiking where bugs might be a nuisance, a hat will help keep them at bay.

### Gloves and Mittens

In mild weather, light wool gloves will keep your hands warm and will not impede finger dexterity. As the temperature drops, though, you're better off with a pair of mittens which will permit maximum circulation. You might think that mittens would be clumsier than gloves, but there is nothing more awkward than cold fingers. Add another layer for extra warmth and protection. Preferably the fabric in your outermitts should allow moisture to escape, but be stiff enough to shield your hands from wind, snow, and rain.

## *Socks and Insoles*

As we mentioned in the chapter on hiking boots, two pairs of socks help to prevent blisters because any rubbing takes place between the socks rather than against your skin. Doubling up on the socks keeps your feet warmer, too, but if you're not careful it can have just the opposite effect. Cramped feet have poor blood circulation and frostbite can result. Before taking the first step, always make sure that you can wiggle your toes.

Wear a thin pair of socks next to your skin. Wool, silk, or polypropylene are good choices for this layer because they do not hold moisture directly against the skin. The catalogs and specialty shops might refer to them as sock "liners," but they're really just plain socks. Next, a thicker pair of socks—preferably wool in cold weather—is worn over these.

Obtaining the optimum fit for the comfort and warmth of your feet is really a matter of experimentation. Try on several different combinations of socks until you find a system that works for you. Don't overlook the possibility of insoles for your hiking boots. They help to cushion your stride and may make the boots fit better. In cold weather they provide extra insulation between your foot and the cold ground. Some hiking boots already come with removable felt liners which can be taken out and dried at night.

## *Gaiters*

Gaiters are nothing more than cylindrical pieces of fabric which cover the lower leg. You wear them right over your pants and they extend over the tops of your boots, thus preventing snow and trail debris from getting inside. They're also good at keeping your pants dry. If you're handy with a sewing machine, you can probably make your own gaiters and save yourself a bundle. I just saw a pair advertised for an outrageous sum. Whether you make your own or buy ready-made gaiters, choose a heavy duty fabric that is wind and water resistant. Rugged material will also hold up better to the abrasive effects of ice, crusty snow, or boots. Finally, make sure your gaiters rise at least fifteen inches up your leg.

## *Booties*

No, I'm not going to advocate sending Baby creeping off into the wilderness. The whole idea behind this product—appropriately named because they do look like giant-sized baby booties—is after-hike comfort. At the end of a day's hike, you remove your clunky leather boots and slip your tired feet into soft, insulated booties. You might say they're kind of like outdoor bedroom slippers. They are really designed for lounging around a campsite, but if you're not camping (which is perfectly acceptable),

*Gaiters protect your pants while trudging through brushy areas.*

booties may still be useful in your alternative lodging. Perhaps you're staying at an alpine hut or an inn. Courtesy prevents you from tramping in with mud-caked hiking boots. Whip out your booties, and presto, you have warm footwear for the evening.

A less expensive solution to the problem might be to carry lightweight running shoes in your pack. Nearly everyone owns a pair these days whether they've ever jogged or not. You don't necessarily have to wait until the day is over to enjoy their comfort. Why not give your feet a break when you reach the summit? Remove your hiking boots and frolic in your running shoes for a few minutes.

## *Bandanas*

You don't have to be a hardened rodeo rider to benefit from the advantages of a bandana. Available in many colors besides the classic red and blue, they can serve the hiker in countless ways. Roll up the square diagonally and tie it around your forehead to act as a headband, preventing salty sweat from dripping into your eyes. Cover your hair kerchief-style or knot the corners for a pirate's hat to keep off sun, sand, or wind. Dab some insect repellent on the bandana and wear it as a neckerchief. This technique will keep the bugs away longer than lotion directly applied to

your perspiring body. In an emergency, you can make a tourniquet by tying a bandana around the injured limb. (Don't forget to release it occasionally.) Need a trail marker? Place your bright bandana at a turn you're likely to miss on the way back and hope the next hiker doesn't take it home for a souvenir. Oh yes, you can even blow your nose.

# Chapter 7

# Nibble your Way to the Peak

## *Hikers' Food*

There's no doubt about it. Puffing and sweating your way up a mountain is hard work. The body's nutritional requirements are intensified, and during a day's hike you might need as many as a thousand calories more than when you are home. Your caloric need is affected by other factors as well. Cold weather hiking places additional energy demands on the body, as does an increase in elevation. Consider this: Walking at two miles per hour, a young adult male weighing between 140 and 145 pounds burns 196 calories per hour over level ground, but 474 calories per hour while hiking up a twenty percent incline. Add a twenty-two pound backpack and that figure jumps to 570 calories per hour, a three-fold increase in energy expenditure.

All this is great news if you're dieting, and if you're not, the course of action seems obvious—eat more. However, good backcountry nutrition is a complex matter, and what you eat and how you eat are just as important as how many calories you stuff into the body. The active hiker needs a constant source of energy throughout the day, supplied in the form of short and long-term energy foods. Marathon-crazed backpackers eat several times a day, rather than two or three big meals. Ingesting large quantities of protein and fat immediately prior to heavy exercise can actually be a liability. Although these foods are important sources of long-term energy, they are hard to digest and will tax the body's metabolism at a time when it is trying to meet a more urgent energy crisis.

Of course individual needs vary depending on one's age, weight, sex, walking speed, etc., and there is no single diet which will be perfect for everyone. But a knowledge of sound nutrition principles will help you form eating patterns which are compatible with hiking. Let's take a brief look at three food types—carbohydrates, proteins, and fats—and see how each relates to the hikers's needs.

## *Carbohydrates*

Carbohydrates are the primary source for your body's heat and energy needs. As opposed to fats and proteins which can also be burned as fuel, carbohydrates relinquish their pent-up power relatively quickly.

There are two types of carbohydrates—simple and complex. Occurring naturally in honey and fruits, the simple carbohydrates provide the quickest energy and are called sugars. They boost your blood's glucose level which must not drop too low or you may feel drowsy, dizzy or confused. Refined sugars fall into this category, but we're beginning to see them as the bad guys because they do not provide any additional nutrients and may actually rob the body of minerals and vitamins required for their own metabolism. They also encourage tooth decay. Stick with the natural sugars as best you can and you'll be a lot better off. Oranges, apples, and peaches are all good sources of natural sugar. Fruits like apricots and apples which come in dried form (or you can dry your own for a substantial saving) are especially suited for trail snacking because they take up considerably less room than their juicier counterparts.

This doesn't mean that a candy bar has no right to be lurking in your pack. In fact, chocolate is one of the hiker's staples. No one can dispute that it provides a quick lift, but remember that your chocolate high is false energy if not backed up by other foods releasing their energy over the long haul. As someone who has pigged-out on the irresistible Swiss chocolates in that country, I can personally testify to this fact. After the initial euphoria wears off, you feel sapped of all strength. Go ahead and indulge once in a while, but don't overdo it.

The second class of carbohydrates—complex carbohydrates—forms a sound basis for meeting the hiker's energy needs. Nutritionists recommend that the average person's diet be composed of as many as fifty percent complex carbohydrates and this proportion should run even higher for hikers. Pasta, bread, potatoes, rice, cereals, and vegetables are examples of foods in this group. Their energy takes slightly longer to be released than the simple carbohydrates because the digestion process must break down their long molecular chains into simple sugars before being absorbed.

## *Proteins*

The body needs proteins like the ocean needs water. Since body tissue *is* protein, we must consume a sufficient amount to build and repair these structures. Proteins are composed of individual elements called amino acids, but no single food source contains all the amino acids which the human body requires. The way to get a complete set is to eat a variety of foods. Red meat, fish, chicken, eggs, dairy products, soybeans, and nuts are protein-rich foods. To a lesser degree, grains, vegetables, and fruits provide protein for the body.

After the stresses of hiking tear down muscles and cells, proteins move to the site to fix the "damage." The process of replacing worn-out tissue requires rest. Although some protein should be eaten with each meal, your biggest intake should be at supper time. During the day a steady stream of carbohydrates fuels your activities, but protein available during the night primes the body for the following day.

## *Fats*

Fats are not something that should be avoided at all costs. They are responsible for many vital functions including the synthesis of hormones, cushioning of organs and transport of certain vitamins. As a source of energy they are real power-charged goodies yielding over twice as many calories per gram as carbohydrates and proteins. Red meat, cheese, eggs, and mayonnaise are loaded with fat. Does this mean you should consume large quantities of these foods before setting out on a hike?

No.

When it comes to fat in the diet, one must consider the *type.* There are fats that are acceptable and others that are probably best to limit. Fats derived from animal sources and those found in eggs and dairy products fall into the better-be-careful category because they are thought to raise the level of cholesterol in the blood, possibly contributing to heart disease. On the other hand, the type of fats known as "polyunsaturated" (referring to their chemical structure) can't be blamed for doing such a terrible thing, and may in fact lower the blood cholesterol. Corn and sunflower oil, and most margarines are examples of polyunsaturated fats.

* * *

There are many books available which tell you how to cook in the backcountry. They give recipes and instructions for preparing tempting feasts. Some of the foods are exotic, some simple, but these books always assume you're a backpacker camping out for several days and are going to haul pots, pans, stove, spices, and freeze-dried this and that up the trail. By the time you eat, you've got enough containers lying around to hold a Tupperware party.

I want to simplify this whole procedure. Let's assume most of your hiking will be of the one day variety. Even if you're hiking in a location away from home for a longer period of time, you don't necessarily have to worry about cooking in the wild. The trick in either situation is to do your preparations at home where you have the facilities to create nutritious meals for both at-home and on-trail consumption.

Start off your day with breakfast in the comfort ot your own kitchen. You were told in grade school that breakfast is the most important meal of the day. This axiom is especially true for the hiker. Without a good supply of energy, you'll be struggling for the first hours of the hike and your strength and morale will probably run in the deficit column the rest of the day.

Once you've been on the trail for a while, pull out the granola bars, dried fruit or trail mix, and nibble your way to the peak. When you reach your destination, celebrate with another light meal, perhaps a hunk of homemade bread with cheese. At this point you should not eat heavily because you must still walk back. If you've kept up your energy level with a proper breakfast and a little snack along the way, you won't feel the ravenous need to consume a huge meal. Another snack in the late afternoon will tide you over until dinner time, when you should eat a big meal.

Which specific foods should you eat? Using the general guidelines given previously for the three food types, put together a combination that will rely heavily on natural foods. As much as possible, include whole-grain products, fruits, and nuts. At first, however, your diet should not differ too much from your normal eating patterns because you don't want to shock your system. Gradually wean yourself away from sugary, salty, or cholesterol-laden food. If you're already eating smart, good for you. You're a natural candidate for hiking and you don't need me telling you what to eat.

Following is a sample menu of the foods a hiker might eat during a typical day. The next section provides several recipes for items which you can make at home and then take on your outing.

***Breakfast***

*1/2 grapefruit*
*Bacon strips—fried crisp*
*2 scrambled eggs*
*Corn-wheat germ muffins (see recipe)*
*1 glass milk*

***On the Trail***

*Late morning:* *Dried apricots*
*Walnuts*
*Lunch:* *Bread with cheese*
*Apple*
*Afternoon:* *Granola bar*

***Supper***

*Tossed salad*
*Pan-fried fish of the season*
*Baked potato*
*Home grown fresh broccoli*
*Rice pudding*

# *Snacks for Hikers*

## *Corn-Wheat Germ Muffins*

*1 1/2 cups flour*
*1/3 cup cornmeal*
*1/4 cup wheat germ*
*3 teaspoons baking powder*
*1/2 teaspoon salt*
*1/4 teaspoon ground cloves*
*3/4 cup milk*
*1/3 cup honey*
*1/3 cup vegetable oil*
*1 egg*
*1/2 cup sunflower seeds*

Combine flour, cornmeal, wheat germ, baking powder, salt and cloves in a large bowl. Add the remaining ingredients and mix with a fork until the flour mixture is moistened. Drop batter into greased muffin pan cups and bake 15 to 20 minutes at 400°F. Makes one dozen muffins.

## *Oatmeal Bread*

*2 packages dry yeast*
*3/4 cup warm water*
*2 cups hot milk*
*1/4 cup shortening*
*1/3 cup honey*
*2 teaspoons salt*
*2 cups flour*
*2 cups oatmeal*

Dissolve yeast in warm water. Mix together milk, shortening, honey and salt. When cool, stir in flour, yeast and oatmeal. Add additional flour as needed to make a soft dough. Knead 8-10 minutes on floured surface. Form a ball and place in greased bowl, turning over once to coat the surface. Let rise in a warm place.

After one hour the dough should have doubled in size. Punch it down and allow to rest 10 minutes. Divide the dough in half and roll each section into a 9 × 15 inch rectangle. Roll these up, squeeze the seams together and place each in a greased loaf pan. Let rise again for 45 minutes. Then bake at 375°F. for 45 minutes. Makes two delicious loaves.

## *Onion Crunchers*

*1 1/2 cups sweet Spanish onion, finely chopped*
*1 package dry yeast*
*3/4 cup warm water*
*2 1/2 cups biscuit mix*
*1/4 cup melted butter or margarine*
*1 cup cornmeal*
*1 teaspoon salt*

Dissolve yeast in warm water and beat in biscuit mix. Knead on a floured surface until even consistency. Form four equal parts and roll each into an eight-inch square. Cut each square diagonally into four triangles. Sprinkle onions on top. Starting with the wide end of each triangle, roll dough tightly toward point forming a long stick.

Mix cornmeal and salt together. Dip sticks first in butter, then in cornmeal mixture. Arrange on baking sheet and let rise for one hour in warm place. Bake at 400°F. for 15 minutes. Turn oven off and allow sticks to crispen for 15 minutes. Makes 16 onion crunchers ideal for trail snacking.

## *Crunchy Crumble*

This unusual mixture has a texture that is neither smooth nor stiff. When you get the hungries, simply crumble off a piece and pop it into your mouth.

*1 cup chunky peanut butter*
*1/2 cup nonfat dry milk*
*1/4 cup 100% bran cereal*
*1/4 cup sesame seeds*
*1/8 cup wheat germ*
*1/2 cup raisins*
*1/4 cup honey*

Place all ingredients in a small, deep bowl and mix thoroughly with a wooden spoon. Use a chopping action to blend loose ingredients into the peanut butter. Pack the mixture into a one-pound plastic margarine container. You can eat as little or as much as you want at one time, and save the rest by snapping the lid back on.

### *Yodel Yippers*

These tasty little fudge cookies require no oven baking. You can whip up a batch for your next outing in just a few minutes.

*1 1/2 cups oatmeal (not instant)*
*1/4 cup peanut butter*
*1/4 cup milk*
*1/2 teaspoon vanilla*
*1 cup sugar*
*1/4 cup cocoa*
*1/4 cup butter or margarine*

Combine oatmeal and peanut butter in a bowl until you have a coarse mixture. Heat the other ingredients gently in a saucepan, stirring constantly. Allow it to boil one minute, then remove from heat.

Mix this liquid mixture into the oatmeal and peanut butter. Use a teaspoon and knife to drop spoonfuls onto a sheet of wax paper. Your Yodel Yippers are ready as soon as they cool. Makes about two dozen.

### *Valley Mix*

Here's an all-natural snack that can be a lot cheaper than pre-packaged fruit-and-nut mixes. Make your own generous servings of this trail mix.

Suggested Ingredients:

*Raisins*
*Dried apricots*
*Pumpkin seeds*
*Walnuts*
*Toasted soy nuts*
*Cashews*
*Sunflower seeds*

Add the ingredients in approximately equal amounts. Use unsalted nuts. Toss lightly. Omit or substitute ingredients to tailor the mix to your own tastes and budget.

## Drink Up

Although we could survive several weeks without food, we would only last a few days without water. Fluid is essential for body functions because it acts as a medium in which the basic life processes are carried on. In addition, water's ability to regulate body temperature by evaporation through the skin and lungs is particularly relevant to the hiker. If too much water is lost without replacement, performance drop and other more serious problems such as heat stroke might result.

*A belt bottle allows access to water without digging into your pack.*

During the course of a normal day, the body needs to take in approximately 48 to 64 ounces of water. This works out to about six to eight cups. Other beverages, fruits or soups can help meet the quota. An active hiker, however, could sweat up a storm and lose 64 ounces of water in a single hour. Therefore his fluid intake needs to be considerably greater to offset the loss.

Even on short, easy trips, it's a good idea to take some water or fruit juice with you because you don't realize how much moisture you're losing. Drink small amounts frequently rather than waiting until you've built an unquenchable thirst. For a change of pace, try squeezing a little lemon juice into the water. A teaspoon of concentrated fruit juice will also perk up the flavor. Available in most supermarkets, one quart plastic bottles are convenient for both carrying and drinking. However, unless you want a soggy pack, make sure your bottle is leakproof before taking it on the trail. Some of the mail-order places sell better quality bottles that are more apt to be genuinely leakproof. Specifically, I've had good luck with the *Nalgene* brand bottles.

One of the nice things about a plastic bottle is that you can freeze water right inside it. Why would you want to do this? When the weather reports promise a scorcher coming up, place the three-fourths filled bottle into the freezer the evening before your hike. The warmth of the following day will slowly melt the block so that you have a constant supply of ice water on the trail.

We've all experienced those muggy days when we've drunk enough water to make the desert bloom, yet still feel thirsty. Sometimes the mouth feels dry even though the body has plenty of fluid. Try sucking on a hard candy. The action will stimulate the salivary glands and alleviate your dry-mouth syndrome.

Many sport enthusiasts worry about losing body salts during heavy exercise. Sodium is an important mineral because it helps maintain a state of equilibrium in the cells, blood and urine. Proper muscle function also depends on this chemical's presence, and a depletion of the sodium reserve can bring on the painful spasms known as heat cramps. The treatment for this condition (which usually occurs in the arms and legs) is to eat salty foods, drink plenty of water, and massage the cramped muscles.

With all this in mind, is it a good idea for the hiker to routinely increase his salt intake? Most doctors feel the action is not necessary and may do more harm than good by raising the blood pressure. If you make strenuous hikes in very hot places where you sweat a lot, you can take in extra salt or even take salt tablets. If you do, make sure you increase your fluid consumption as well.

When you explore the backcountry longer than a single day, you might not be able to carry enough water to meet your needs. Even on a short hike, that cool sparkling stream looks awfully inviting after a hard climb. But is it safe to drink the water?

In a brochure prepared for visitors to Alaska, the Arctic Health Research Center reminds hikers that the clarity of the water is not a good indicator of the presence or absence of potentially harmful microorganisms. Many clear streams give a false sense of security. Since conditions change daily, public health officials can't guarantee the safety of any particular stream.

Unfortunately, with wilderness areas more heavily visited than ever, the incidence of contaminated water is on the rise. The single organism which is causing the most concern is *Giardia lamblia.* According to the Federal Center for Disease Control in Atlanta, it's the most common intestinal pathogen in this country today. As many as sixteen million Americans may harbor *Giardia.*

Existing as a dormant cyst in surface water (streams, lakes, etc.), the parasite goes wild once it enters a human host. The organism clings to the digestive tract where it multiplies and causes such symptoms as diarrhea, stomach cramps, bloating, gas, and loss of weight. These symptoms appear one to three weeks after *Giardia* is swallowed. It is absolutely necessary to get medical care in order to kill the parasite and be cured.

The best way to avoid *Giardia* and other harmful contaminants is to play it safe. Assume that all surface water contains deadly organisms and take the necessary steps to kill them. There are several ways you can accomplish this. The first sterilization method is boiling. I know it sounds rather impractical to stop in the middle of the day and boil large quantities of water, but who wouldn't enjoy lingering a few moments over a steaming cup of herbal tea on a chilly slope? The major boiling chores are suited for the evening hours after the day's activities because you can ready your water supply for the next day's hike. You'll need a portable backpacking stove and generous supply of fuel. Although three minutes at a rolling boil is long enough to render *Giardia* harmless, twenty minutes must be allowed in order to assume that all other organisms have been killed (at higher elevations, five and thirty minutes, respectively). You might find the water has a rather flat taste after it has cooled off. Replace the lost air (the flavor) by pouring the water back and forth between clean containers two or three times.

Disinfection by chemical means is somewhat more fussy than boiling, but eliminates the extra weight of stove and fuel. Add four drops of household bleach such as *Clorox* to a quart of water and shake well. Loosen the cap and shake again. This second shaking insures that untreated water in the screw cap threads will come in contact with the disinfectant. Then let the water sit for thirty minutes, but double the contact time for cloudy or cold (less than 41°F.) water. Don't add fruit juice or other flavorings until the waiting period has passed because they might inhibit the disinfecting process.

Iodine is well known for its antimicrobial properties. However, excessive amounts taken internally can be lethal. The commercially available brands of hydroperiodide eliminate the guesswork by providing pre-measured doses of iodine in tablet form. You'll find the product carrying such brand names

*You can easily boil water for tea or coffee with a portable backpacking stove.*

as *Globaline* and *Potable-Aqua.* Simply drop the tablets in water, shake until dissolved and start timing.

A new method of sterilization that holds some promise is filtration. Scientific laboratories routinely use filtration setups for separating out microorganisms. Unfortunately hikers and backpackers have not yet been able to buy an inexpensive filter that is practical in the field. The situation may be changing, however, as some manufacturers adapt science to the general consumer's needs.

Some people who would never dream of slurping water from a stream, commit an equally dangerous act. They "drink" snow. Reasoning that nature's newfallen layer is pure and untainted by human activity, they quench their thirst by melting snow in their mouths. The danger does not really lie in contamination but stems from a basic chemical principle. A great deal of energy is necessary to transform snow at 32°F. into water at 32°F., even though no rise in temperature occurs. If the process is fueled by the body's metabolism, then it becomes a heat-robbing operation and undermines a body already threatened by cold and thirst. Carry a small backpacking stove on cold, snowy hikes where you can't take enough water. Not only will the stove serve to change snow into water, it will sterilize the water as well.

Day hikers who would like a hot trail drink don't have to miss out if they lack a stove. Coffee, tea or hot chocolate can easily be carried in a *Thermos* bottle. To prolong the time that the contents will remain warm, pre-heat the *Thermos* at home with boiling water before adding the hot beverage.

You might be tempted to pack some alcoholic refreshment for your trip and I've certainly got nothing against a little nip now and then. In fact what could be more romantic than a champagne brunch in an alpine meadow dotted with spring wildflowers? However, beware the use of liquor during cold weather. Yes, you will initially feel warmer after a drink because the effect that alcohol has is to dilate the peripheral blood vessels. At a time when the body's normal response is to restrict circulation to the hands and feet in an effort to conserve heat, the unchecked flow of blood to these areas will cause heat loss. Better leave the schnapps at home or transport it unopened to your mountain cabin where you can enjoy it in front of a crackling fire. And never give alcohol to a hypothermia or frostbite victim.

Chapter 8

# Finding Your Way

## *Routing Your Trip*

This chapter contains what I feel is the essence of the book's philosophy. You don't have to be a hard-core backpacker to enjoy hiking. Nor do you even have to go to the mountains. Every state has accessible hiking trails which will lead you to solitude.

What constitutes a trail?

In the United States there is a system of marked paths which criss-cross our wilderness areas. Unfortunately, the pattern of ownership is not unified and the land falls under the responsibilities of various federal, state, local and private groups. Whatever their political boundaries, the trails are open to everyone and I have never paid a cent to walk on public ground. Parking is another matter.

Since land management does not come under a single authority, there is no centralized source of information about the trails. A wealth of maps, brochures, and guidebooks exists out there, much of it free, but you must scout out the literature yourself. One of the purposes of this book is to assist you in locating appropriate material for enjoyable hikes.

Sometimes booklets may be available at the site of the trail as is usually the case with nature trails like Five Rivers Environmental Education Center in Delmar, New York or Creamer's Nature Path in Fairbanks, Alaska. These are short loops with numbered signposts at ecological points of interest—perfect for a leisurely stroll on a Sunday afternoon. The path is usually in pretty good condition—relatively smooth and wide—without

*Don't rely on memory for directions. Take the map with you!*

long, steep climbs. You won't need full-fledged hiking boots, but do wear something comfortable on the feet.

Another source of information about trails is your state government. There are literally thousands of state preserves and parks. Many agencies put out leaflets describing selected trails and you shouldn't feel bashful about utilizing the resources that your tax dollars have financed. Check the appendices at the back of this book.

A bookstore or backpacking shop in the geographic area where you wish to hike is also a good bet for finding guidebooks to backcountry hiking trails. Most of the larger bookstores in and around Albany, New York carry several books on the nearby Adirondack Mountains. Travel to New Hampshire and the situation is the same for the White Mountains. In the larger cities like New York you'll be able to find trekking guides covering most of the world. Another good source for a variety of reliable guidebooks for hiking abroad is Bradt Enterprises, 409 Beacon Street, Boston, MA 02115.

If you already know which specific area you'll be hiking in, a topographic map can be a useful tool in navigating your course. Ask for them by quadrant. Sometimes you can find them sold locally, but they can always be ordered directly from the government. For information about ordering and the cost of each map write to the following addresses. Request a free index to the state(s) which interests you and then check which quadrant your trail falls under.

*United States*

| | |
|---|---|
| All states west of the Mississippi River | Branch of Distribution<br>Central Region<br>U.S. Geological Survey<br>Box 25286<br>Denver Federal Center<br>Denver, CO 80225<br>(303) 234-3832 |
| States east of the Mississippi River, including Minnesota | Branch of Distribution<br>Eastern Region<br>U.S. Geological Survey<br>1200 South Eads Street<br>Arlington, VA 22202<br>(703) 557-2751 |

*Canada*

| | |
|---|---|
| 3 free indexes cover entire country | Canada Map Office<br>Department of Energy,<br>Mines and Resources<br>615 Booth Street<br>Ottawa, Ontario<br>K1A OE9<br>(613) 998-9900 |

In case you've never seen a topographic map before, you're in for a treat. Besides being useful in the field, the maps are actually quite attractive and make impressive decor on any wall. A friend recently bought one as a souvenir *after* we hiked the area.

Topographic maps depict the contour of the land so you can tell whether you will be hiking uphill or downhill, and how steep the grade is. Sometimes actual trails are marked on the map. At first glance the map looks like just a bunch of wavy lines, but it's not really difficult to interpret them. Each continuous line represents an elevation (in feet or meters) and the figure is printed at least once somewhere along its length. When you walk in the same direction that the line runs, you are walking at a constant elevation or level ground. A route that traverses the lines at an angle indicates a climb or descent, and the closer the lines are spaced, the steeper the land. If you hold the map at arm's length from your face, the lines take on a three-dimensional appearance which will help you visualize elevations with different colors for a greater life-like quality. To learn more about map reading, pick up a copy of Kellstrom's ***Map and Compass*** which is probably the best book on the subject. Good skills in relating a map to the terrain will pay dividends later, possibly in preventing you from getting lost.

The many hiking clubs and organizations (some of the major ones such as the American Hiking Society are listed in the appendix) are frequently overlooked sources of maps, books, pamphlets and helpful newsletters. Some clubs have formed separate publishing companies (Sierra Club Books and Appalachian Mountain Club Books, to name two) which print their own guidebooks. Even a simple mimeographed newsletter might alert you to an obscure, but pleasant, hiking trail. Outdoor clubs are always sponsoring group activities, and you'll benefit in several ways from hitching up with a hiking party. You'll meet interesting people, learn from the expertise of veteran hikers, and find good trails. Word of mouth is ultimately the best publicity any place can receive.

Once you get started, you'll find more trails than you can hike in a lifetime. Aside from the simple beachcombing or guided walkway type of hike, the trails you'll encounter most frequently will be marked, maintained trails. This means that some organization has taken responsibility for erecting permanent markers to identify the correct route whenever it is not self evident. The markers can be simple metal tags nailed to the trees at periodic intervals to assure the hiker that he is still on the right path, or an actual sign at a confusing junction. Whatever the method of identification, the best markers are those which are the most environmentally compatible. On some barren sections of trail above timberline, for example, you may encounter a stack of rocks, known as a cairn, indicating where the trail is. The unobtrusive symbol creates a much more pleasing image than a garish fluorescent orange arrow painted on a boulder.

In addition to working out your route, you should also be concerned with an estimate of time. You wouldn't want to start on a trail, only to

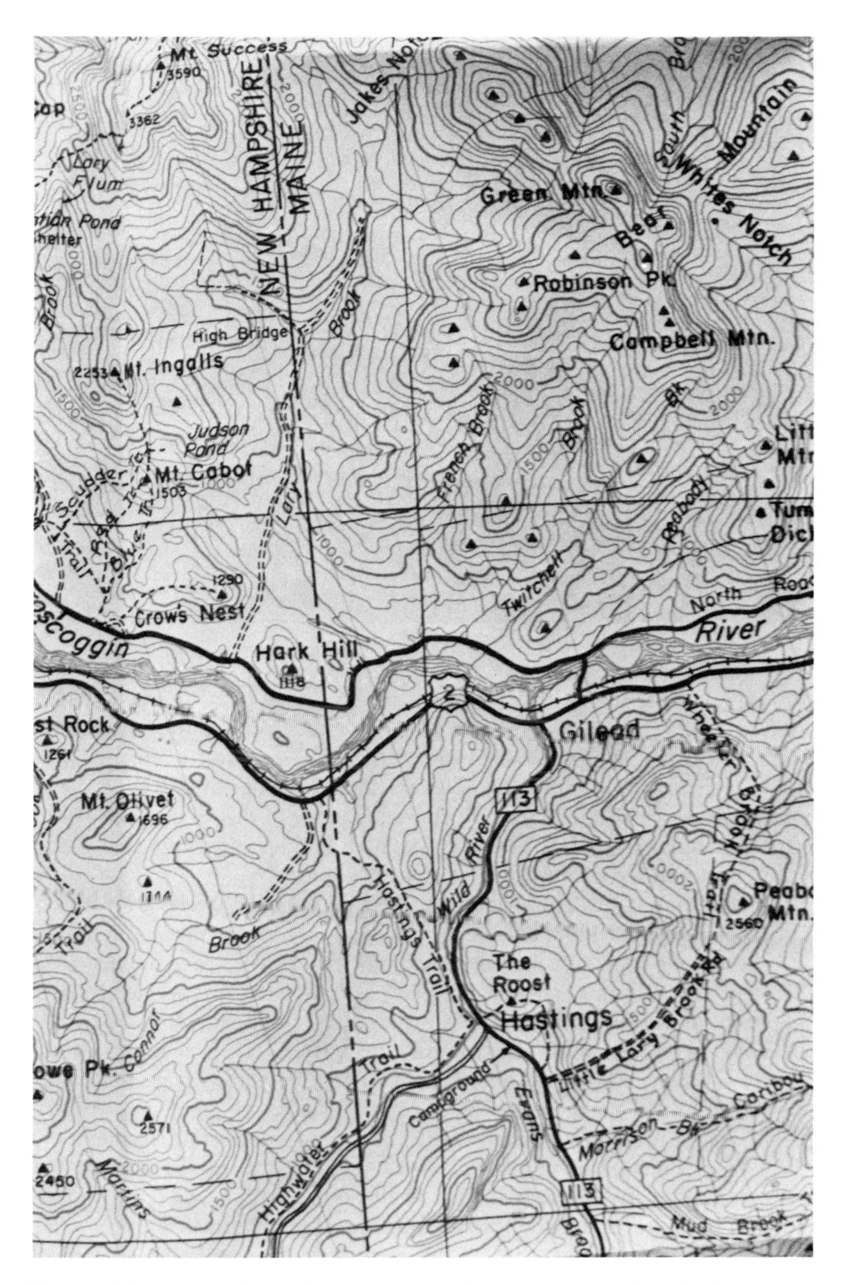


*Topographic maps can be useful because they show the contour of the land. (© 1979 Appalachian Mountain Club)*

*A "cairn" above timberline.*

discover that you had not allotted enough time to make the return trip before darkness. Although it is safe to conclude that the further you venture, the longer it will take, you can not gauge your hiking time solely on the basis of distance. As any runner-hiker will tell you, a two-mile neighborhood jog sounds like a piece of cake, but tackling that same distance over hilly terrain is another story.

A better indicator of the time involved in any hike is elevation, with a minimum of one hour required for each 1000-foot vertical increment. Of course your time will vary depending on your own pace, but let's use this figure in a sample calculation. Suppose you plan to conquer Mount Tippytop, elevation 3500 feet. You drive to the trailhead located at 1000 feet above sea level and begin your hike from there. This means you must climb 2500 feet. Allowing one hour per 1000 feet, you will need two and one-half hours to make the ascent.

Although the return trip is downhill, the going can be slow. Remember, it's not really any easier coming down than it is going up. To play on the safe side, allow another two and one-half hours. So now we're up to five hours. Throw in some time for smelling the flowers and listening to the birds and you can easily see that this particular hike will require a good portion of your day. If you aren't willing to devote this much time, then you should choose a simpler route. Look for trails that are not as long and do not encompass such a steep rise.

# Accommodations

Despite the most generous schedule, there is something anti-climatic about turning around and retracing your steps when you've reached your destination. Hiking enthusiasts who have graduated from novice status will want to take longer hikes and perhaps venture further than a single day allows. Does this mean the old tent routine is inevitable?

No. There are a variety of backpacking alternatives which provide opportunity for exploring wilderness areas, yet satisfy your need for dry overnight shelter. One of the highlights of a trip to Alaska with a friend from college days was our hike along a damp, mossy path to Laughton Glacier. At the end of the trail we entered a comfortable, though primitive cabin equipped with plywood bunks and a wood heating stove.

Had such facilities not been available, we probably would not have made the hike. The structure enabled us to sleep soundly that night, and the following morning we made the short side trip to the base of the glacier itself where we marveled at the sun glinting off the ice. The Forest Service of the United States Department of Agriculture maintains more than 130 such cabins throughout Southeastern Alaska within the area designated as the Tongass National Forest. The cabins are accessible by trail, boat or float plane. My friend Steve and I reached the Laughton Glacier trail by a narrow-gauge railroad, which unfortunately has gone out of business, but I'm sure a weary hiker would find the other units in the system just as suitable. The $10 per night fee charged for the use of the cabins may easily be the best bargain you'll find in Alaska these days. If all this sounds like your cup of tea, write for information and reservations to:

USDA Forest Service
Attention: Pamela
P.O. Box 2097
Juneau, AK 99803
(907) 789-3111 or 586-7151

In response to your request, you'll receive a packet of information that includes a pamphlet on the Tongass National Forest, an application for a cabin permit, and a list of suggested supplies for your stay in the cabin.

Mountain "huts" in New Hampshire are similar to the cabins in function, but differ in other respects. Designed to serve more hikers than just your own party, the huts provide communal sleeping arrangements in two or more bunkrooms equipped with mattresses, blankets, and pillows. A resident crew serves breakfast and dinner at fixed hours. The real beauty of the hut network is that the individual buildings are located a day's hike apart from each other, and with shelter and meals provided, all you need to carry are personal items. William E. Reifsnyder's ***High Huts of the White Mountains*** is a comprehensive book of hut lore as well as a good trail guide to this picturesque area of North America. Hut-hopping is also

*Zealand Falls Hut (elevation 2700 feet) in New Hampshire's White Mountains.*

popular in the Austrian and Swiss Alps. In fact, this is where the whole idea got started. To request a pamphlet describing the hut system in New Hampshire and its costs write:

Appalachian Mountain Club
Pinkham Notch Camp
P.O. Box 298
Gorham, NH 03581
(603) 466-2727

You need not be a member of AMC to stay in the huts, but reservations are recommended, especially for the summer months which sell out very early. Hikers without reservations may stay if space allows.

Within some of the national parks you can find an assortment of lodges and cabins run by private concessioners. Personally, I find them a bit overpriced, but their chief advantage is that they lie in the heart of the park grounds close to the trails you wish to explore. This saves you travel time in and out of the park. If the national park is one which does have such facilities, you will find them mentioned in the park's general pamphlet with an address to write for further information.

You probably already know about youth hostels if you're a teenager or young adult in your early twenties. But if you don't fall into this category, did you know that youth hostels are open to everyone regardless of age? The purpose of American Youth Hostels, Inc. is to provide opportunities for travel and outdoor recreation for all. Quite extensive in Europe, the youth hostel system has been expanding rapidly in the United States and Canada.

What can a hosteler expect when he or she checks in?

You will receive simple overnight accommodations (translation: bed) in a room with other members of the same sex. Other than that, there is great variation among individual establishments which can range from grungy to posh. Some hostels are very small and homey and have kitchen facilities where you can cook your own food. Other larger buildings are more dormitory-like and may serve meals. Whatever the physical characteristics of the place, I think you'll find one important aspect that is uniform. The atmosphere is genuinely friendly and conducive to conversation. I've met people from all over the world staying in hostels.

When you look at the costs of motels and hotels these days, hosteling is dirt cheap by comparison. Sure, you're not getting all the conveniences, and a steady dose of non-privacy can frazzle your nerves, but the savings you gain by staying in hostels once in a while could make the difference between going on the trip or not. In order to stay at a hostel, you'll need a membership card. The fee for the card is less than the price of one night's lodging in a motel and is valid for the whole calendar year. For further information write:

American Youth Hostels, Inc.
National Administrative Offices
1332 "I" Street NW, 8th Floor
Washington, DC 20005
(202) 783-6161

Everyone likes to be pampered now and then, especially the hiker, and what could sound better than waking from a good night's rest under a cozy quilt to the aroma of homemade oat cakes and coffee? If you think such luxury belongs to another era, consider one of the fastest-growing tourist phenomena today: bed and breakfast.

This option probably won't save you too much money over the cost of commercial lodging, but if you've been anesthetized by an overdose of bland motels and fast food, it will definitely save your sanity. The concept is aimed at lending a personal touch to your travel. You will be staying in someone's home, quite possibly in a room vacated by grown children or college-bound offspring. You and other guests may even be invited to join your hosts in their living quarters for an informal evening of conversation or music before you retire to the privacy of your own room.

The next morning you'll wake refreshed and receive a big country breakfast to send you on your way. Although bed and breakfasts cater to the traveler in general, their nature makes them especially suited for the hiker. Many are beginning to crop up in prime hiking territory. Since the concept of B&Bs (as they're called in England) is relatively new in this country, listings of homes are somewhat divergent. There are several guide books now on the market (see appendix) which might make good starting points in helping you locate a bed and breakfast to suit your needs, but these books shouldn't be considered as definitive listings since their subject matter is still evolving. Do check with local Chambers of Commerce and tourist agencies because they can help put you in touch with the B&Bs in their area. A welcome development in several regions is the creation of central service numbers which you can call to make reservations.

Although they are increasing rapidly, bed and breakfasts in the United States are not yet plentiful or close enough to plan a hiking loop of several days' duration without the use of a car between points. I'm sure this will change in the future.

When it comes to providing the ultimate level of creature comforts for the hiker, inn-to-inn hiking through New England's mountains and countryside goes one step beyond the bed and breakfast concept. While staying at historic inns along various hiking routes—Vermont's Long Trail, for example—you will eat gourmet meals. Several package tours advertised in the travel pages of newspapers and magazines offer this kind of activity on an organized basis. The tour operator even transports your luggage to the next hotel on your itinerary; so all you have to do is walk. I've never been on one of these guided vacations myself, but for the beginner it might be the perfect introduction to hiking because it removes the stress of finding accommodations and restaurants.

If you think all of these backpacking alternatives sound rather self-indulgent, remember that they may actually help you enjoy ***more*** of the wilderness experience by freeing you from the time-consuming camping chores of set-up, cooking, clean-up, and personal hygiene. When you allow yourself to guiltlessly enjoy the outdoors by day and the indoors by night, hiking becomes a stand-alone activity that is indeed pure and simple.

## *Setting up a Hiking Log*

I've always been a filer, clipper, jotter, and planner. You know the type. The pain-in-the-rear kind of guy who straightens his desk every five minutes and has all his tax records on a home computer. One of the reasons you go hiking is to get away from these sick people, not become more like them.

But wait. Before you dismiss this section entirely, I'd like to show you how a little organization can pay dividends.

If you make a habit of carrying a notebook with you on hikes, you can keep a daily journal of important data. A small three by five inch notebook is more than adequate; the blank pages of anything larger are too intimidating. Some of the entries you'll want to make are: date, location, weather, time begun, time completed, distance and vertical rise.

In a previous section, *Routing Your Trip,* we lightly touched upon how to estimate your hiking time. An accurate log can be an invaluable asset in determining more precisely the time it will take you to make future trips. Although many guide books will give you the "average" hiking times for the trails they describe, nobody is an average hiker. You can use these figures, however, as rough guidelines and combine them with the data in your log book to come up with a better estimate.

Go on several hikes at an unhurried pace. For each trip, record your starting and finishing time and then compute the difference. Compare the total elapsed times of your various hikes with the quoted times. If your own times vary consistently—either higher or lower—then you can work out a correction factor with which you can adjust the book estimates.

Another useful parameter is vertical rise. The guide book might list this figure, or you can calculate it from the beginning and ending elevations of the trail as shown on a topographic map. Use this number and your actual elapsed hiking time to come up with your rate of ascent. Since the huffers' and puffers' pace on steep trails will vary considerably from that of the speed demons, a knowledge of climbing rate can therefore be more indicative of the total time required for you to tackle a mountainous route.

A hiking trip has a way of bringing out the photographer in all of us. You become so inspired by the grandeur of the landscape that you can easily expose several rolls of film in no time at all. When all the pictures come back, however, it's not so easy to distinguish them from each other. Although the idea of making notes about the pictures you snap is really nothing new, adopting the structured format of a log book will make the chore seem less cumbersome.

You already have the date, time, and name of the trail filled in. In one column list the frame numbers; in a second column give a descriptive word or phrase if it applies, such as "Button Point" or "Roger's Cove." When you look at the pictures a year later, they will mean much more if you can associate a name with a particular scene. For the meticulous information chronicler, a technique which will help pinpoint the site is to write the frame reference number on the trail map itself.

Make sure you leave room in your log book for trail observations. No lengthy dissertation is necessary, but do state any pertinent or unusual conditions which will be of future value. For example, suppose you are hiking on an exposed mountain ridge in Maine. You note that the fall foliage has passed its peak color and many of the trees have already shed their leaves. Next year when Uncle Stanley again invites you to use his mountain hideaway, you can check your log book. You will know enough to schedule your vacation a bit earlier to take better advantage of the autumn colors.

Is the trail muddy or dusty? Is your favorite wild flower in bloom? Does the trail pass through a field of ragweed which gets you sneezing? Do the bugs swarm up from a nearby marsh and eat you alive? Factors like these are rarely revealed in the guidebooks, but if you have them recorded in your log, several options open up. Your schedule can avoid certain hikes or take advantage of particular conditions for other hikes. You may wish to avoid the area altogether.

Why not reserve a space in your log to record your own personal "enjoyment index"? After you have completed the hike, rate it on a scale of one to ten. This figure will provide an easy method of comparing hiking areas long after your memory has lost its objectivity.

# Chapter 9

# Special North American Hikes

You'll find mountains, seashores, and canyons. Swamps, deserts, and prairies. Our continent is a vast land of contrasting landscapes which offer many opportunities for unique hikes. Whatever type of climate and terrain you prefer, you can find a hiking area to suit your tastes.

By definition, hikers are a breed of individualists seeking out their own kinds of experiences. They're greedy. Never content with one experience, they must move on. Fortunately, this need is easy to satisfy. Here are some suggestions for the adventurer in you. I've chosen several of my favorite areas across the United States and Canada.

## *Alaska*

Although it's been said and written so many times to the point of sounding trite, Alaska *is* an immense wilderness and you really can't appreciate its vastness until you've visited this great land and the neighboring Yukon Territory of Canada. I spent an entire summer with a friend touring the region and we covered only a fraction of the total land mass.

Spanning four time zones, Alaska has glaciers that are larger than some states.

Hike through an abandoned gold rush settlement in Kluane, a newly created Canadian park facility. In Alaska itself, Denali National Park which contains 20,300 foot Mount McKinley, the highest peak in North America, provides hikers with their own personal escort service. The free shuttle buses traveling the sometimes treacherous dirt roads will deliver you to anywhere within the park and will pick you up when you're done. Use this opportunity to track an elusive Dall sheep over a craggy mountainside. When you see one of these wandering white specks in the distance, hop out of the bus and go for a hike. You'll be able to approach very close without the animal spotting you. Evolutionary selection has favored the development of a keen sense of horizontal sight at the expense of vertical visual acuity. The sheep are easily able to spot predators such as wolves, but are less adept at seeing humans coming toward them.

I could go on forever about the abundant wildlife in Alaska, but too many descriptions begin to sound exaggerated. For example, Steve and I used a guidebook on our trip which kept touting the splendors of this or that sight along our driving route. One location in particular was supposed to feature a river brimming with spawning salmon. A bit suspicious, but afraid we might miss something good, we stopped the truck to check out the claim. I reached the bank first, and indeed there were dozens of brilliant red salmon fighting and jumping their way upstream. Wanting to share my enthusiasm for the bonanza, I called to Steve but his response was a cool, "Yeah, sure." Only when he saw the spectacle himself, did he become a believer. He was astounded by the awesomeness of the sight. Alaska in general elicits the same reaction from the first-time visitor.

## *California*

### *Lassen Volcano*

Lying at the southern tip of the Cascade range, Lassen Volcano in California offers a rare perspective for the intrepid hiker who is willing to undertake a strenuous climb. Standing at 10,000 feet inside the lava dome (inactive, of course), you witness the devastation caused by the earth's explosive power. Further evidence of volcanic action exists on the unearthly Bumpass Hell trail which takes you past a number of hot springs, steam vents, mud pots and boiling pools. These hydrothermal areas are particularly eerie when a fresh snowfall is on adjacent ground.

If your definition of natural beauty follows more traditional lines, other hiking trails have abundant vistas of forests, sparkling lakes, and colorful wildflowers. The national park is also a habitat for many species of wildlife including mammals, birds, amphibians, and reptiles.

## *Death Valley*

Everyone wants to be a comic these days. While I was visiting Death Valley, California, I placed a phone call to New York City. The person I wished to speak with was not in the office, but the secretary asked if I would leave a number where I could be reached. Since I was traveling and really did not have a number, I hesitated a moment, not knowing quite what to say. The operator quickly jumped in. "Well, you know, he's stuck in the middle of the desert."

Although I wasn't stranded as the operator's comment implied, hiking in Death Valley does give one a sense of the isolation early prospectors must have experienced when seeking a short-cut to the gold fields through one of the harshest environments on earth. Their nearly fatal experience gave the valley its name.

The time of year should be your first consideration when planning a trip to Death Valley. An all-time high temperature for North America was recorded in July 1913 when the floor of the valley reached 134°F. Temperatures are usually above 100°F. in the summer. The most comfortable hiking weather occurs between October and April, but avoid holiday periods as they're the most crowded.

Within Death Valley lies the lowest point in the Western hemisphere—282 feet below sea level. You'll be pleasantly surprised to encounter a great deal of geological diversity within the park. Hike among sand dunes, dry lakes, strange mineral formations, colorful badlands, and even a volcanic crater.

# *Arizona*

## *The Grand Canyon*

What can one say about the Grand Canyon other than yes, it sure is? You could stare at it for hours and still not comprehend its reality, let alone understand the complex geological processes which shaped it millions of years ago (and still play an active role today). John Muir, the famous naturalist, described the canyon's surreal beauty in 1898, " . . . as unearthly in the color and grandeur and quantity of architecture as if you had found it after death on some other star."

Although every tourist should allow plenty of time when visiting the canyon, this is especially important for the hiker. You will be compelled to pause at many points along your driving and hiking routes. I had the notion before I visited the Grand Canyon that there was only one canyon to see. Not true. There are many side canyons which offer their own special cliff-side vistas to help you appreciate the Grand Canyon as a whole.

While the size and diversity of the canyon make for intriguing sightseeing, they do complicate the logistics of exploration. There are two

"rims" to the Grand Canyon—north and south. Geographically, they are separated by only ten miles, but you must invest a day's driving time to get from one to another because you obviously cannot drive across the canyon. One rim is not "better" than the other. Both are worth a visit.

I suggest that you first drive the park roads along either the north or the south rim, doing a general "survey" of the area. Then go back to do some hiking on one of the designated trails. You can repeate this plan of attack on the opposite rim.

Unless you have made reservations for one of the park lodges or have an overnight camping permit, make sure you are back at the trailhead before sunset. You wouldn't want to miss nature's show as the lights dim and the canyon walls take on a red glow.

# *Utah*

## *Zion and Bryce National Parks*

If the Grand Canyon isn't enough to inspire a little wonderment, cross the border from Arizona into neighboring Utah where Zion and Bryce national parks await your visit. In Zion, you can hike among the towering cliffs where huge bedrock masses have been tossed about like children's blocks. Your trip through Bryce Canyon will be characterized by such names as Inspiration Point and Fairyland View. A little imagination will easily transform the intricately sculpted rock formations into castle turrets, pagodas, Turkish soldiers, or many other uncommon features.

# *Florida*

## *The Everglades*

If you'd prefer a less arid hiking experience than the desert offers, how about a swamp hike? Florida's Everglades provide the opportunity and you don't necessarily have to get your feet wet. Thoughtful park planners have constructed boardwalk trails in many areas of ecological interest to promote the theme that this land is not just a big swamp. The natural complexity of its terrestrial and aquatic plant and animal communities is skillfully revealed by trails which take you through representative "life zones."

Hike along a trail where stately pine trees grow right out of dissolved limestone potholes. Or marvel at some of the largest mahogany trees in the continental United States. You've never seen as many different kinds of grass as from the Pa-hay-okee Trail where over one hundred species flourish in the marshy glades from which the park gets its name.

Home of both migratory and resident birds, the Everglades is an ornithological paradise. Long time and neophyte bird watchers alike will delight in the antics of the anhinga, a fish-eater who flips its catch into the

*Boardwalk hike through Florida's Everglades is a civilized way to observe swamp ecology.*

air before swallowing it whole. Other species include the great white heron, egret, roseate spoonbill (often mistaken for a flamingo) and brown pelican.

Your hike through Florida swampland would not be complete without a glimpse of the American alligator. Once threatened with extinction by poachers who profited from the demand for alligator shoes and handbags, the reptile is now making a strong comeback.

The best season to visit the Everglades is during the winter. Hiking can be uncomfortable from June through September when the hot, rainy season brings out the bugs.

## *Nova Scotia*

If Florida is too hot for you, try traveling Northeast all the way to Nova Scotia. Nova Scotia brings to mind quaint fishing villages along rocky coastlines. But the northernmost part of the province has a landscape that is reminiscent of the Scottish highlands. Little known as a hiking destination, Cape Breton Highlands National Park sits atop a 2000-foot plateau which drops sharply to the sea along its western edge. The lush forested cliffs are a sharp contrast to the sparse subarctic vegetation of the interior plateau where lichens and mosses cling tenaciously to exposed rocks.

Whether you want to pick blueberries in high-altitude isolation or gaze out to sea on a windswept promontory, there is a hiking trail which will lead you there.

In addition to the impressive scenery, hikers in Nova Scotia enjoy another important advantage. The bed and breakfast system is at its finest, with charming homes and cordial hosts who take pleasure in helping you get the most from your visit. Indeed, the attitude of the general population seems to be one of respect for the outdoor lover. "Nova Scotians don't think of backpackers as people who just use the land and then leave litter behind," claims Bill Bryson, Outdoor Promotion Supervisor at Tourism Nova Scotia. "They understand that backpackers probably respect the land as much as, or even *more* than most everybody else, and leave it as clean—maybe cleaner—than before they got here. Throughout Nova Scotia people understand the need to get off and live simply on foot or on bicycle."

Bryson also offers the suggestion that visitors go on a highway hike. Simply treat the many scenic roads which hug the 4600 miles of coastline as you would a trail. No one will find your action the least bit unusual.

# Chapter 10

# Trail Techniques

## *Learning to Walk Again*

Anyone who can walk, can hike. However, hiking does place additional demands on the body and proper technique will help you ease the stress that your heart, lungs, back, feet, and legs must bear.

Pay particular attention to basic body mechanics as you hike. If you shuffle through daily life with sloppy posture and a haphazard gait, you'll pay dearly for these habits on the trail. Point your toes forward in the direction that you are walking to avoid the unnatural ankle torque which results from an inward or outward orientation of the foot. Although actual pain from slight twisting would not be immediately noticeable, the effect is cumulative and you might suffer discomfort several days later. And if you repeatedly hike with a turned-in or turned-out foot, you increase your chances of inducing long-term physical damage.

Avoid fast movements on uneven ground and never jump or leap when carrying a pack. Your leg muscles are already bearing extra weight, and the sudden stress of even a short hop down a step could rupture the calf muscle or achilles tendon. You'll feel as if someone shot you in the leg.

While hiking over level terrain, try to carry the torso as vertically as possible. A fully erect posture distributes your body weight in the most efficient way and is especially important if you are carrying a pack. On uphill climbs, your initial tendency will be to lean far forward, but this creates strain which will tire you more quickly. Resist the temptation to

*Where would you step? Stepping around the rock uses less energy than stepping up, then down.*

overcompensate for the inclined angle. Lean only slightly forward on uphill slopes.

Downhill hiking is tricky. A steep grade is murder on the knees which act as your brakes against the forces of gravity and momentum. If you lean backward, you only increase the pressure and risk the possibility of causing the equal and opposite reaction—feet lurching forward, right from under you. If your knees begin to ache, bend slightly *forward* from the waist as you move downhill. Although the position will initially seem very awkward, you'll soon get the hang of it. I have a favorite step which I resort to when downhilling becomes too oppressive. I make sure the path ahead is clear; then I turn around so that I am facing uphill and take a few backward steps. This momentarily eases the pressure on the knees as well as stretches the calf muscles. I do this a few times during the course of the descent.

Placing your footsteps slowly and deliberately can be beneficial to the hiker in two ways. First, you prevent yourself from stumbling or twisting an ankle. Second, if you step over small obstacles such as logs and rocks, you perform less work than if you were to step up, and then down. When you tally the day's physical expenditures, the net result is less fatigue.

And fatigue is something that all hikers must combat. Many beginners have the queer impression that they must hurry to the end of the trail to earn their "reward" of a lofty vantage point. The sooner they get there, they reason, the sooner their agony will end and they will be able to rest. Such psychology is not only self-defeating, it is based on false assumptions.

You will not get there any faster by racing uphill. After several minutes of this punishment, your heart will be pounding and you will need to stop to catch your breath. Eventually you'll be too pooped to plod. So what have you gained?

The best pace for hiking is an easy, slow gait of about one and a half to two miles per hour. You'll knock off a ten mile hike before you know it and arrive at your destination fresh and cool. Keep the number of stops to a minimum because starting up after a complete stop is harder on the body. Try to keep moving, no matter how slowly. Sitting down slows blood circulation and allows warmed muscles to stiffen. If you must rest, lean against a tree or rock (but don't sit) every half hour or so. You'll have more stamina for the long haul.

The "lockstep" is a technique which experienced hikers use that actually allows them to rest *while* they're moving uphill. How is this possible? They introduce a slight pause into their gait, thus creating a moment when no physical exertion is needed.

Make the lockstep work for you on steep climbs. When you place the forward foot upon the ground, straighten the rear leg to a full standing position. With knee locked, rest on this stationary leg long enough for an actual pause, but not so long that you break the stride. Then repeat on the other leg. The process may seem needlessly slow, but will conserve precious energy on vertical rises.

Proper breathing technique is also important to hiking efficiency. Your goal is to maximize lung capacity. Quick, shallow breaths do not draw enough oxygen into the body. Take deep, long breaths as you climb uphill, making sure to inhale and exhale fully. This action will permit as many cubic inches of air as possible to fill the lungs.

Should you carry a walking stick? The very notion conjures up visions of a hundred-year-old Indian guru hobbling to the summit of a Himalayan peak. But a hiking staff can be a practical aid in certain situations. Use it to lean on when you are negotiating a rock-strewn slope or tiptoeing across a stream. The stick acts as a third leg to catch your weight if you happen to lose your balance. I just saw a carved walnut hiking staff with engraved brass plate advertised for $30 in one of the equipment catalogs. You can take a less bourgeois approach and simply pick up a stick along the way and discard it when you're finished. Or, you could whittle your own fancier model.

*Lean against a rock to rest. Sitting decreases blood circulation and allows muscles to stiffen.*

*The lockstep: Pause briefly with weight resting on "locked" rear leg.*

*Use a walking stick as a third leg on uneven terrain.*

# Hiking Etiquette

With the increased use of wilderness areas, noise pollution is becoming quite a concern. Most hikers go into the wilderness to enjoy the calm, relaxing atmosphere. They don't want to hear a noisy crowd. It's so easy to get carried away and forget that other people may be nearby. There's no need to whisper or creep stealthily; just carry on conversation in normal tones, especially if you know other people to be in the same area. Ear-splitting cries are fine at a soccer match, but they don't belong in the wilderness.

Use courtesy when you encounter other hikers face to face as well. Step to the side for descending hikers. While hiking in a group, the person in the lead sets a steady pace. No one should feel rushed. The outing is a pleasure trip, not a race. If the person in the rear must run to catch up, then the pace is too fast. Besides, sprinting up a slope and then stopping is the least efficient way to hike. This method overworks the heart and lungs and is actually more tiring than climbing without rest periods.

Even if you know the area, don't wander from the group. The others may have to wait for you to return. If you don't take exactly the same route back, you risk the possibility of missing them altogether. A good rule of thumb is to always keep a member of the group in sight. If you need to stop or absolutely must leave the group, be sure to tell someone.

The lure of wilderness areas presents a paradox to those of us who treasure them. We want to admire and touch the beauty, but when too many people do this the land is no longer a wilderness. The "minimum impact ethic" is a principle which states that one should always take the course of action which endangers the land in the least possible way, or has the "minimum impact."

Suppose you are choosing a place to hike this weekend. You can apply the concept of minimum impact by avoiding a popular spot in favor of a less-traveled area. The trail to Mount Marcy in New York State's six million acre Adirondack Park is a sad example of an overused trail. Torn sod and trampled flowers scar the mountainside and in some places the trail itself is fifty feet wide. Water runs through muddy gullies during wet weather.

What caused this?

Thousands of cleated boots dislodged the soil and nature simply could not keep up with the repairs. The area was overworked. Officials at Denali National Park in Alaska are trying to prevent a problem like this from occurring. Overnight backpackers must obtain a permit before setting out. By limiting the number of people in certain zones, rangers hope to protect the more fragile and popular areas.

As an individual you can lessen the impact of your presence on the land. Seriously consider off-season use. If heavy snows have not arrived, a January trek over a frozen trail can be a delightful experience on a cold, sunny day. Your lug soles will not tear the ground as they would in the summer.

Of course you will want to do some hiking during the warm weather, too. On these occasions you can still apply the minimum impact ethic. Leave the landscape exactly as you found it. Whatever you carry into the wilderness, you should carry out. Why not take a large plastic garbage bag with you and help remove litter left by less thoughtful hikers?

Do not, however, take anything that is part of the natural environment. The item may seem insignificant to you, but everything plays a role in nature's scheme. For example, scientists are discovering that dead twigs may be very important to the ecology of the desert area where wood is scarce. Even an abandoned bird's nest might be reinhabited next year.

Authorities in certain national park areas with historical significance also like to stress the Federal Antiquities Act of 1906 which protects artifacts such as arrowheads, bottles and tools. These implements are often the only remnants of a past culture. Removing the artifact or dislodging it from its location makes it difficult to recreate the past. Yosemite National Park in California contains artifacts from prehistoric civilizations nearly 5,000 years old. And the Chilkoot Trail in Alaska is a living monument to the thousands who risked their lives—and sometimes lost them—in their pursuit of the Klondike goldfields in 1898. The trail is littered with the abandoned equipment of gold seekers.

If you are hiking on a marked trail, stay on the trail and don't make short cuts. They create unnecessary erosion. Never construct your own markers because they only confuse others.

Above all, love the wilderness. It is yours to enjoy. Every person who uses this precious gift shares the responsibility for its protection.

## *The Handicapped Hiker*

The wilderness experience is something that does not need to be forsaken by the handicapped, and in fact may have special therapeutic value for the physically disabled as well as those with psychological problems. Because the activity is non-competitive, the handicapped individual can feel good about himself. He does not need to pit his abilities against others in a no-win situation. He is out there pulling his own strings.

Athletes who have become disabled as a result of sports injuries may gain a special psychological benefit from hiking. The "displaced athlete" is able to transfer his athletic yearnings to a new activity. Thus hiking becomes a surrogate sport which minimizes the trauma of the original loss.

The achievements that some disabled persons have made in this field are outstanding and their level of fitness would put many of us to shame. The next time you grumble when faced with a long uphill hike, think of Larry Clark, a six foot six inch, 200 pound man who enjoys backpacking in the woods of Ohio. The characteristic that makes Clark different from the normal weekend sportsman is that he does his hiking in a wheelchair.

Paralyzed from the waist down as the result of breaking his back in a car accident, Clark acquired his outdoor skills at the Vinland National Center in Loretto, Minnesota where medical director Dr. Keith W. Sehnert conducts a self-care training program. The program goes beyond the usual rehabilitation process of coping with day-to-day living and promotes activities which enable the disabled person to lead an independent, assertive lifestyle.

Larry Clark, who holds a full-time job as a watch repairman, has found that his physical revival has given him more energy and helped him to manage stress better. "Rehabilitation is teaching you how to live," he states. "I learned I can accomplish anything if I try. I'm a better person now."

I've noticed that our country is now much more aware of the needs of the disabled than it was ten years ago. Every wheelchair-bound individual may not feel like tackling rocky, rutty terrain, but this doesn't necessarily make him or her an improbable candidate for a hike. Simply choose trails which you can negotiate easily in a chair. Look for wide trails with gentle inclines. Many localities such as Lafayette, California have nicely graded bicycle or jogging trails that are ideal for leisurely wheelchair strolls.

Several units of the National Park Service have upgraded their facilities and services to enable handicapped and wheelchair-bound visitors to enjoy more fully the sights and sounds of the parks. There's no longer any excuse for leaving Grandma in the car while everyone else rushes over to the breathtaking precipice. Park rangers conduct daily activities suitable for anyone. She might, for example, enjoy participating in a "geology walk" through Yosemite Valley where she can enjoy the scenery and contemplate the geologic history of the area. When planning your trip to a national park, write and ask for information about wheelchair trails, or inquire at the visitor station. In addition, certain park roads traveled exclusively by shuttle buses will allow private vehicles displaying the universal wheelchair emblem. Ask.

Also ask about the Golden Access Passport. This certificate allows blind and permanently disabled persons free lifetime entrance to all of the national parks, monuments and recreation areas. Everyone riding in the car also gets in free, and there is a fifty percent discount on fees for such things as camping, boat launching and parking. The government does not yet have a centralized system for issuing the Golden Access Passport, so you should directly contact the area you plan to visit.

I'd like to mention the Golden Age Passport for persons sixty-two and older. The Golden Age Passport grants the bearer free lifetime entrance to park areas administered by the federal government. You can't get the pass by mail, but otherwise the process is relatively simple. Just show up with proof of age at any of the national parks. Even if you just *look* old, you might squeak by a kindly ranger. While visiting the Everglades National Park with my parents this year, I encouraged my father to get a Golden Age Passport. We stopped at the visitor center before the park entrance, but they were all out of passes. At the entrance gate itself, the attendant did not bother checking proof before allowing us to drive through.

My parents are healthy and do lots of walking, but our excursion through the swamps revealed what an excellent opportunity the boardwalks provide for less ambulatory people. You can easily push a wheelchair along the well-planned paths. Developers of beach parks are also catching on to the idea of boardwalks which help to make this terrain accessible to the handicapped. It's also encouraging to see the boardwalk concept carried out in other less tropical climates. If our society is to develop a genuine appreciation of the natural environment, we must extend the opportunities for enjoying it to every citizen.

# Chapter 11

# Trail Delights

## *The Hiking Photographer*

Hiking and photography go hand in hand. Many people are enjoying 35mm photography nowadays and there is no better place for it than on the trail. There is a vast amount of material on the subject. Entire books can, and have been written on photography. My intent, however, is not to expound upon the entire body of photographic knowledge. Instead, I'll just present a brief collection of personal tips to help you get better pictures.

If you are about to buy your first 35mm camera, I urge you to get a manual model. "Manual" means you must set the aperture (the size of the lens opening) and the shutter speed yourself. All the advertising and advice of friends will try to convince you otherwise, and believe me, you'll be tempted by the sleek automatics which promise to do everything—and do. That's the problem. For the beginner they execute too many functions. A camera that automatically computes the exposure and makes the settings, never allows the owner to learn firsthand the interrelationships between light, lens aperture and shutter speed, and the effects various combinations produce. When you let the camera make these critical decisions for you without a working knowledge of its limitations, you're not using it effectively as a tool. You've surrendered creative control to an object. This is fine if your only objective is to get middle-of-the-road

*Photography is a good way to help children gain an appreciation for the wonders of the outdoors.*

pictures. But I suspect that the person who goes to the trouble and expense of pursuing this hobby wants more from his efforts than mediocrity.

True, most automatics do have an override feature that converts the optic system into manual operation. But if you're just starting out, you'll tend to favor the automatic mode. Most people who blissfully learn "photography" on automatics don't know how to shoot pictures by making the manual adjustments themselves. There's no question that it's hard work and time-consuming. Your initial attempts will be very frustrating. It took me a year of practice on my manual camera and lots of film before I was able to consistently get results with which I was pleased.

If you're still not sold on the idea of a manual and you can't pass up an irresistable buy on an automatic, make a vow to use the camera in the manual mode for several weeks before switching over to automatic operation. The skills you acquire during your training period are well worth the small investment of time.

The next measure I'm going to advocate also goes against the grain of popular opinion. I thought I was the only person in the world who held this view, but I just read an article by an outdoor photographer who said the same thing, so now I won't feel as if I'm committing heresy when I suggest you *not* buy a skylight filter. The traditional thinking on this matter is that the filter protects your lens and corrects for hazy atmospheric conditions. However, any extra piece of glass you place in front of your lens increases the possibility of distorting the light that must eventually reach the film. I believe such filters may reduce the overall sharpness of your pictures. If you are careful while using your camera and keep a lens cap on at all other times, there is no reason to fear lens damage.

The situations you encounter which present the most scenic photo opportunities are often the least likely places you can stop and manipulate your equipment to take a picture. The method of carrying your camera on the trail can make the difference between capturing the fond memories of a hike on film and a hassle-filled juggling session. Most cameras come with a thin plastic strap. Worn around the neck, the strap can become quite painful as the camera itself bounces and tugs. Before you take your camera on any long trips, remove the strap and replace it with one of those wide cloth bands. If you plan on doing a lot of hiking and want your camera poised at all times, you might even consider a chest strap which holds the camera closer to your body and prevents it from flapping against you.

For those times when you do not actually want the camera around your neck, you can carry it in a day pack. Manufacturers are coming out with some clever camera packs which keep your gear organized and ready for use. The packs assume a variety of shapes and styles, and the case that is right for you depends a lot on personal preference as well as what equipment you wish to place inside. All packs are not necessarily slung over the back or shoulders, either. A friend of mine has a very convenient waist pack. No one has ever accused Julianne of passing up a chance to shoot a photo, and this kind of pack allows her to be quick on the draw.

If you're going to spend money on a camera pack for trail use, consider several features. Look for water-resistant material with reinforced seams. The inside of the pack should be well padded. Take your camera, film, and accessories right to the store and put them inside the pack. If the items are wedged tight, you will not be able to retrieve them easily, but they should not be so loose that they roll around. With the pack fully loaded, try it on and make sure you can move freely without any binding or discomfort.

Regardless of the type of pack you use to carry your camera, there is one additional piece of material you should always use during cold-weather photography. Remember to place your camera inside a plastic bag before returning indoors to prevent condensation on electronic parts.

Save time on the trail by anticipating the types of pictures you intend to snap. Excessive changing of lenses in the wilderness is not only inconvenient, but creates a greater risk of exposing delicate interior camera components to dust and moisture. A hiker who plans to photograph expansive mountain ranges should start the day with a wide angle lens on his camera, not wait until he is standing at the edge of a cliff. An interest in wildlife photography, on the other hand, would warrant a telephoto lens. My sister likes to hike with an adjustable focal-length lens of 28-

70mm. At the lower end of the scale she has wide angle capabilities, but can "zoom" in on a subject by increasing the setting.

In dimly lit situations such as a dense forest you might wish you had a tripod to steady your camera, but who wants to lug the monster on a hike? Thanks to modern technology, you don't have to give up the idea. Consider one of the tiny collapsible models whose legs measure four inches or so and weighs only a few ounces. You set the device on the ground or a rock, or even fasten it to a tree with a *velcro* strap. The secure footing enables you to use slow shutter speeds without fear of shaking the camera.

If you like experimenting with new techniques, there's a gimmick on the market now called a split-field filter which solves the age-old squabble of whether you should focus on the mountain or the wildflower. Depth of field limitations had previously forced you to make a choice between one or the other. If you wanted the foreground in sharp focus you had to put up with a fuzzy background, and vice versa. No longer. The split-field filter is like having bifocal glasses for your camera. The top half is clear, allowing unaltered distance perception, while the bottom half is a close-up lens.

There's another type of split-field filter which can be useful in outdoor photography. One half is gray and the other half is clear. With the gray half positioned on top, the filter prevents overexposure of the sky while permitting details of the darker ground to show clearly.

Don't think you're "cheating" by taking advantage of these little tricks. Professional photographers use them all the time to enhance natural features of their subjects and make them more visually appealing. Here's a well known technique: Carry a small spray bottle on your hike. For close-up shots of flowers, leaves, spider webs, etc., simulate a dewy morning by gently misting the surface. Beads of moisture will catch the light, adding interesting highlights to your picture.

If you were to take the same picture at another time of day, the end product might look different. The angle at which light strikes the subject varies as the sun travels across the sky. The best times for photography are usually in the early morning and late evening when the low incident angle of light accents detail. The composition of the light itself is also different during these periods. The light contains more reds and oranges and will cast a warm glow on the scene. I'm sure you're aware of the colorful skies during sunrises and sunsets, but the effect can be quite dramatic on sand, snow or rock.

"Backlighting" is a term used to describe the situation where light comes from behind the subject toward the camera. With a little box camera, I grew up thinking this was a bad thing, to be avoided at all times. I now realize backlighting can be a valid photographic technique when used properly. Sun-dappled autumn leaves are downright gorgeous when the light passes through them. You must be careful, however, with backlit backgrounds because opaque objects may appear too dark if you've set your exposure for the background. Don't be afraid to use a flash in the great

*A dewy morning or photographer's trick? Only you and Mother Nature know for sure.*

*Use the sky as a dramatic backdrop for a silhouette shot. (Courtesy Karen A. Drotar)*

outdoors. For example, in the autumn scene above, suppose you wish to photograph Aunt Kitty hiking through the forest. Adjust your exposure for the foliage, focus on your aunt, and snap. The light emitted from the flash will illuminate Aunt Kitty but will have no effect on the distant trees.

The use of a camera in the backcountry allows you to capture a special scene and look back at a later time. Take some extra time to set up your shot and adjust the camera. But don't get so involved with the photography that you forget to participate in the moment. The best photographs can never match the memories of the actual experience.

## *The Trail Ecologist*

The air is dewy fresh with the scent of spruce. Hiking alone through the pristine forest, you haven't a care in the world. Yet somehow you realize you're not really alone. You're a part of a greater totality we call the environment. Although hiking can be a pleasant experience in itself, your awareness of some simple ecology can make the activity even more meaningful.

Ecology is the interrelationship of organisms and their environment. Every plant and animal on this planet depends on its surroundings, and sometimes the web of interdependencies produces ingenious adaptations for survival. The further you delve into the biological significance of specific phenomena, the more amazed you'll be at the great powers of endurance found in nature.

The wild rhododendron in the Smoky Mountains is a good example of an organism which has evolved protective mechanisms. During the blooming season, the stems of the flowers produce a sticky substance to discourage ants from climbing up and raiding the pollen. Flying insects, however, have free access to the cache of goodies and are able to spread the pollen to other flowers, thus assuring continuation of the species through cross-fertilization.

In the winter months, you might see the rhododendron's leaves drooping pitifully. They aren't dying. The rhododendron is an evergreen plant, and to slow down the rate of moisture loss from the undersides of its leaves, the plant curls them more and more tightly as the air becomes colder and drier.

Virtually every hike you make presents the chance to learn about a new ecological tidbit and observe it in action. A good place to start is a state-operated nature preserve where conservation biologists have prepared educational displays and programs to help visitors understand the characteristics of that area. At the Oak Orchard Wildlife Management Area near Batavia, New York, you can observe via television hookups a program known as "hacking" which trains baby eagles transplanted from Alaska to live in the wild. The baby birds, who are only eight weeks old, sit in cages atop towers. Keepers supply the birds with food, but all humans remain hidden from view. After four weeks, the cages are opened and the birds are free to go in and out, gaining valuable flying and hunting experience. Each day the food is reduced and eventually the eagles are able to survive on their own. State officials hope the bald eagle population will increase so that the species is no longer endangered.

I'm a big fan of the programs sponsored by the national parks. Pick up leaflets and check the bulletin boards for information about ranger-guided hikes and campfire programs addressing a variety of topics from migratory birds to snakes. The presentations are always well-researched, informative, and entertaining.

Once you've graduated from these programs, you'll see that there are many ecological activities which you can conduct on your own. Have you ever keyed a tree? It's fun and a lot more satisfying than a video game. The purpose of the activity is to identify which species a tree belongs to. Just imagine the impression you'll make on your friends when you mosey up to an oak tree and say "Gee, Fred, isn't that a *Quercus palustris?*"

The process of keying involves a botanical reference manual whose information is divided and subdivided according to characteristics of the species. For example, if the tree has pointed leaves, you turn to one part of the book; if the leaves are rounded, you turn to the other part. Then you check another characteristic, such as bark. Smooth bark, flip here; rough bark, flip there. If you know anything about computers, the whole idea is the same as a binary search. In fact, keying is so foolproof that you end up with the answer staring you in the face. Then you can read more about what you have found.

Budding biologists who wish to participate more actively in their quest for ecological insights can now test rain, snow, lakes and streams for unusually high acidity levels. Pocket Lab, a kit marketed by Early Winters, Ltd., weighs only one ounce and measures the acidity and alkalinity of water. The problem of acid rain has received much coverage in the media. Formed when atmospheric moisture comes in contact with sulfur dioxide emissions from coal-burning industries, acid precipitation is fatal to fish and other aquatic life. There are now over two hundred fishless lakes in New York's Adirondack region as a direct result of acid rain.

Perhaps your interest does not lie in the acquisition of data. That's okay. After all, who says you have to run around like a mad scientist? You can still enjoy seeing and learning about the affairs of Mother Nature. The more you understand her ways, the more this knowledge will manifest itself in appreciation and a preservation ethic.

Here are a few more "Gee whiz" kinds of facts that I have come across in my own hiking. I hope they inspire you to do some exploration on your own.

## *Avalanche Chutes*

Densely forested mountain slopes will sometimes exhibit a vertical swath of destruction caused by the ruthless fury of an avalanche which crushed everything in its path. Although it is sad to see so many beautiful trees annihilated, the action plays an important role in nature's scheme. New underbrush springs up and provides a source of food for foraging vegetarians such as moose who would otherwise have a difficult time finding food in the area.

## *Permafrost*

If I had to choose one byword from my trip to Alaska, permafrost would be it. In the arctic region, the ground freezes solid for most of the year. During the short summer, the surface thaws to a depth of several inches and whatever plant life exists must survive in this fragile zone. The underlying region which remains permanently frozen year-round is called permafrost.

Permafrost is responsible for some pretty weird stuff. A forest of spruce trees can appear drunken with tipsy trees leaning every which way. Since the soil layer is so thin, tree roots do not penetrate deeply and are very susceptible to the heaving process of seasonal freezing and thawing.

*Tree roots cannot penetrate the permafrost layer in this sparse Alaskan forest.*

In tundra areas north of Fairbanks, the annual rainfall averages only four inches per year. Yet the terrain is spongy and water oozes up around the hiker's boots. Why? Permafrost prevents rain and melting snow from soaking into the earth. Instead the water forms small pools and streams. The standing water is prime breeding territory for some of the most ferocious mosquitoes I have ever encountered.

## *Preserving your Hikes*

Here I go again. I can never leave well enough alone. In a previous chapter, I advocated a hiking log for recording data on the trail, and now I'm going to expand upon the idea. Whether you're journalistically inclined or just want to tap some of your creative juices, you can preserve the hike in a memory book of photos, descriptive passages and even poetry.

Unlike the log book in which you simply recorded facts, a more comprehensive memory book will enable you to capture moods. Every place evokes a feeling. Are you comfortable here or do you get a feeling of uneasiness? Sit down right on the spot with pen and pad and spill your guts, as if you were talking to an old friend. Tell about your physical surroundings—sights, sounds, smells. What thoughts do they stir up? Try to relate the environment to something personal in your life, but don't worry about literary perfection. Just write. Unless you choose otherwise, no one has to see the composition except yourself and it makes no difference if your sentences are a bit awkward.

Now take your camera and aim it at a scene which is representative of the landscape. Snap. Record the frame number in your notes. When you get the roll of film developed, retrieve the picture and your essay of the heart. Arrange them in one of those photo albums where the plastic pages cling to the underlying surface and can be lifted up to insert items.

Here's a sample entry derived from my own wanderings. See the accompanying photo on the opposite page. I chose to write the description in the third person, but you can use any style that seems comfortable to you.

*Photo number 16*
*5-13-82*
*West side Toggenburg valley, Switzerland, above Wattwil*

*He climbed the steep hill above the village and reached an open meadow. Sitting on the sloping lawn, he watched a cloud poke its head into an otherwise cloudless sky. It hesitated, faltered on the horizon not quite daring to interrupt the perfection of the day. Beneath it, dark pines shaded a small culvert wedged between two meeting slopes. Sunlight filtered onto a few trees beginning to leaf out. In the foreground he saw richer, greener color in the grass occasionally dotted with pieces of white flowers which had only recently replaced the white of last month's snow. The air felt mild as puffs of breeze gathered strength and spit at him. The sun beating on him was even warmer.*

*Photo from "memory book."*

*Here one escaped the traffic of the village into an idealistic world. He felt as if he had entered an earlier era. Cowbells clattered on the hill behind him. The frozen landscape of winter had thawed into a rushing, soothing brook punctuated by a twittering bird. From the valley floor, church bell melodies punctually floated upward every hour, a subtle reminder of the time-structured Swiss society surrounding him. Spring had arrived on schedule, the grass was growing according to plans and even the cows were now making milk on daylight savings time. Exhausting their supply of grass, they would soon be sent away to summer camp on yet higher mountain slopes. The cloud disappeared beneath the horizon and it became clear to him. He, too, had fulfilled his purpose here and would move on to a new meadow.*

Maybe the piece won't be a candidate for the next Pulitzer prize, but the whole idea here is that I have something concrete I can look back on. Words stir emotions that are sometimes lacking in a picture.

Of course, you don't have to adhere to the formula above to create your own brand of *belles lettres*. If you enjoy poetry, a tranquil lookout might be just the place to spend the afternoon. Here's a poem my sister wrote during her travels.

*The Mountains*

*You're all a beauty to behold.*
*Your magnificence can not in words be told.*
*As a light covering of clouds go wisping by,*
*Your face so stark and strong*
*Is silhouetted again against the sky.*
*You evoke a myriad of feelings*
*Amongst men of all kind.*
*To some, awe and fear,*
*To others peace of mind.*
*What is it about you*
*That makes one want to stare,*
*To gaze hour upon hour from afar*
*And to forget each and every care?*
*Is it your size, your strength, your beauty*
*That everyone wants to see?*
*Or simply to marvel at such a creation*
*The Lord has brought to be?*

Maybe you're not a writer. In that case, don't rule out sound recordings. The availability of smaller and better-quality tape recorders facilitates self-expression. And here's another possibility: trail videotaping. Compact new equipment permits trail movies for playback on your television set. With the whole outdoors as your stage, you can sing and dance your way to stardom. Even if the idea of singing strikes you as preposterous (I wouldn't be caught dead doing it myself), I can't think of a better way to capture the uninhibited performances of children.

Let your imagination run wild. Whether your tools are pen and paper or advanced microelectronic technology, hiking under open skies can spawn the creativity that is buried within you.

# Chapter 12

# Hiker Awareness

## *Safety First*

I deliberately saved this chapter until last because I was afraid I might discourage some borderline hikers by warning of all the misfortunes that can happen to them. Hiking is not a particularly dangerous sport, but you shouldn't undertake it blindly. Wise planning and anticipation of a variety of conditions are the best safeguards a hiker has against the dangers of the wilderness.

Going with an experienced companion on your first few outings will do wonders to ease any fears you might initially harbor. Always tell someone where you are going and when you expect to be back. If you have no one to leave word with, let a ranger know. Sometimes a trailhead may have a sign-in sheet on which you indicate your destination and the time you started hiking. Making use of the log is to your advantage should you become lost.

Although you will probably check the weather forecast and dress for the conditions, be prepared for any unpredicted weather changes. Also plan your hike according to your fitness level and keep within your limits. The chances of an accident are greatest when the body is over-exerted.

When traveling in remote areas, use caution crossing streams. The water level can fluctuate as a result of melting snow at higher elevations. The trickling stream you cross in the morning may turn into a raging torrent as warm afternoon temperatures cause it to rise. Likewise, you should be aware that flash flooding can occur in desert areas, particularly in late summer. If the sky looks at all threatening, stay out of narrow canyons. They could become dangerous stream beds fed by rain falling miles away.

If you do decide to ford a large stream, take a few minutes to study the conditions. How deep is the water and how fast is it moving? Choose your point of crossing as carefully as you chose the trail. The widest part of a river is usually the safest because the current is slower there.

Wear sneakers or leave your boots on, and join hands with other hikers in your group. If you cross in teams of two or three, other members can counteract one person's loss of balance. Face upstream, but move downstream at a slight angle, testing each step before you place your full weight onto it. Never try to move directly against a strong current. If you get scared at any point, don't hesitate to go back. You accomplish nothing by pushing yourself beyond your level of confidence. Look for another place to cross.

Even the most conscientious hiker who has taken all the proper safety precautions can encounter emergency situations in the wilderness. When misfortune strikes, the most important point to remember is to stay calm. If a member of your group becomes ill or injured, stop immediately. Administer whatever first aid is necessary to stabilize the victim's con-

*Would you be trapped if a flash flood hit this dry creek bed?*

dition. Make him comfortable and then seek help, always leaving one person with the injured.

Carefully note the following details and go to the closest state, local or federal law enforcement office for aid.

Name, address, age and sex of victim
What happened
Victim's condition
Exact location

Don't move the victim until help arrives because you might create additional injury. The obvious exception to this rule would be if a greater danger exists by remaining where you are. In that case, use utmost care in moving the injured, and if help is delayed, construct an emergency shelter. The hollows behind large rocks or tree trunks will give protection against the wind.

If you become seriously sick or injured yourself with no one nearby for help, take time to analyze your situation and form a plan. Although you may be tempted to shrug off pain, stay where you are and don't try to fight the elements in your weakened physical condition. You need to conserve your energy by keeping warm and quiet. You can live for days without food if you have a water supply and think rationally. Acting out of panic can spell death.

The same calm approach can be the key to your survival if you become lost. Look around and try to remember how you got to that point. Are there any familiar landmarks or trails nearby? If possible, climb to a high point with a good view and survey the landscape, looking for roads, buildings and other signs of civilization.

If your scan turns up nothing, you still have other options. Try to locate a river or creek bed. Following it downstream will usually bring you to a populated area. In the absence of any landmarks or streams, it is better to stay in one place than to wander aimlessly. Choose an open area so that you can be more easily spotted, and lay out a piece of bright clothing as a signal.

If there is no sign of help by late afternoon, plan to spend the night. An emergency flare in your pack could be set off to attract attention. Always pack a small pocket flashlight for your hike. You'll have a better night knowing you can reach for light whenever you need it. In any case, don't wait until the evening grows dim to make preparations for the night because there is nothing more terrifying than stumbling around in the dark. Make yourself comfortable now and you'll greet the next day with clearer thinking.

## *Critters Large and Small*

Whether we like it or not, we must share the outdoors with a multitude of flying, crawling, biting, and stalking creatures. This section will present a few tips for peaceful co-existence. Let's start with the small critters first and work our way up.

Bugs can be a real nuisance as you hike through the woods, but there are ways you can keep them away from you. Stock up on insect repellent and always carry a small bottle in your pack. I've found the most effective brand is *Muskol*. The potion is very expensive (a single ounce of the stuff might run you five bucks), but you only need to use a little bit at a time because it contains one hundred percent active ingredients. Put a drop or two on the back of your hands, and then rub around. Don't waste precious drops on the insides of your hands. If your budget forces you to try a cheaper brand, check the label for the chemical compound diethyl toluamide (DEET). The greater the percentage of this chemical, the more powerful the repellent. Remember that insect repellents are **very** strong chemicals. Avoid getting lotion in the eyes. During storage periods, keep the container itself away from packs and clothing to prevent the disappointment of seeing expensive gear destroyed by the caustic action of leaking fluid.

For reasons which are not yet fully understood, body chemistry may affect the degree to which insects find you appetizing. Scientists have found that people who ingest large quantities of vitamin $B_1$ excrete the excess through their sweat glands and apparently mosquitoes and other little buggers find the scent objectionable. I once tried taking $B_1$ tablets, but I didn't really notice any difference in my susceptibility to insects. However, your own body metabolism may be conducive to this sort of treatment and it might be worth a shot. Some folks say mosquitoes don't like garlic, either.

Don't neglect some simple non-chemical ways of avoiding insects. In really buggy territory, wear a heavy long-sleeved shirt with the sleeves rolled down. If the insect population is especially persistent, you can try to seal them off from you by donning a light-weight parka with drawstring hood and elasticized wrists, and tucking your pants inside your socks. Insects apparently are attracted to the color blue; so you might want to select your outdoor wardrobe with this in mind.

The time of day, season, and weather can all work in your favor (and against the bugs) if you know how these factors relate to insect life. Mosquitoes are generally most active during early morning and twilight hours, so if you're hiking where mosquitoes are a problem, avoid these times. Some pests have an active, but limited, annual season and you're better off if you schedule your hiking trips around these periods. For example, the dreaded Adirondack blackfly brings virtually all outdoor pursuits to a standstill for several weeks in spring when adult flies emerge from flowing streams.

If you hike during sunny, breezy days, you can minimize the effect of insects. Most bugs are grounded by wind speeds greater than five miles per hour. Even on relatively calm days, there may be a breeze blowing across a barren ridge and you can choose your trail to take advantage of such geological features. At the very least, you can move constantly to prevent bugs from lighting.

Although most bugs are merely an annoyance to hikers, some species may transmit diseases. When doctors recently reported a few isolated cases of Black Death thought to be caused by flea bites, I became as paranoid as the day I first learned about cooties. Actually, there's no reason for panic, just precaution. Black Death—or plague—dwells primarily in rodents. A human can catch the disease from direct contact with an infected animal or from the bite of an infected flea. The best prevention therefore, is to avoid rodents and fleas. I assume you don't go around handling rats, but stay clear of mice, chipmunks and squirrels also, no matter how friendly they may seem. While hiking in plague areas (America's southwestern states), apply insect repellent to socks and pant cuffs, and then tuck the pants inside your socks. Leave pets at home because they are an open invitation to fleas.

Ticks and chiggers are other kinds of outdoor pests with ominous-sounding names. They lurk in grass and low vegetation waiting for an animal to pass by. The bite from a tick can spread Rocky Mountain Spotted Fever, a misnomer because the majority of cases now occur in the Atlantic and Southern states. Tick and chigger prevention involves measures similar to those for fleas. Keep shirt tails and pant legs tucked in and apply insect repellent before hiking in infested territory.

Pollen may not be considered a "critter," but it can be a beastly problem for thousands of allergy sufferers. There is good news, however, for hikers who are afraid to go outdoors during the hay fever season. The conditions may not be as bad as you think. Wooded areas usually have very little ragweed, the most common sniffles inducer. Furthermore, the higher you climb, the less likely your chances of encountering offensive allergens because pollen is heavy and tends to sink to the bottom of a valley. You can take advantage of this principle by choosing trails which follow high ridges. If you're a real mountain person you can hike above timberline to escape grass and tree pollens as well. I discovered these marvelous facts while hiking in Switzerland. My hay fever miraculously disappeared when I went hiking in the mountains. Your own hiking may not be so alpine oriented, but do consider higher elevations if you're bothered by allergies.

Desert hiking is another good alternative for allergy victims. There's virtually no pollen flying around because the dry climate does not promote the growth of luxurient pollen-bearing species. Also, the molds associated with damper climates are conspicuously absent in the desert. If you live near an ocean or lake, go to the beach for relief from hay fever. Sea breezes will push contaminants inland.

One of the most widespread allergic reactions is to poison ivy, with nine out of ten people suffering from itching, swelling, and blistering. Growing almost everywhere in the United States, poison ivy can exist as vines, bushes, or individual plants. However, just about all forms produce bright, shiny leaves in clusters of three. If you have doubts about whether a particular plant might be poison ivy, recall the rhyme—"Leaflets three, let it be."

The reaction to poison ivy is caused by the oils which, once touched, tend to spread over the entire body. Wash the affected area with soap and water immediately after you notice itching or redness, or suspect you may have come in contact with the plant. Then apply a hydrocortisone cream, such as *Cortaid,* to soothe the itching. Change your clothes because they probably contain the oils also.

Mushroom connoisseurs who take to the woods may be inclined to stop along a damp, shady path and pick a few sprouting fungi for their dinner. That's fine if you are absolutely sure you can identify them, but a word of caution here, please. I just read about a poor gentleman in New York State who died as a result of eating mushrooms he mistakenly identified as a type which grows in his native Finland. Mushroom look-alikes are so deceptive that you must be able to pinpoint the organism right down to the exact species. For example, the Parasol Mushroom (*Lepiota*) is considered a delicacy, but Morgan's Lepiota, a member of the same genus, is poisonous. Some people assume that certain mushrooms could not be poisonous if wild animals eat them, but this reasoning is flawed. Although rabbits can safely nibble *Amanitas,* this mushroom will kill a human. The only safe rule is to refrain from eating wild mushrooms unless you are sure the species is edible. As fungi, the spores of mushrooms cross-pollinate. You should be careful of even "safe" mushrooms.

Wilderness animals pose special problems for the hiker. Within the last several years, the incidence of rabies has doubled. Although many creatures look temptingly cute, never try to feed or handle a wild animal. The fleeting moment of pleasure is not worth the risk of getting bitten. Seek medical attention immediately if you ever do get bitten by an animal. In addition, human interaction is not good for the animals themselves. These words from the *Yosemite Guide* explain the hidden dangers:

"By feeding or petting a wild animal, you help make that animal less wild. Many deer and bears have abandoned berries and leaves and insects to seek human food . . . And just because they eat it doesn't mean it's good for them."

"Some animals fed and petted by humans become unpredictable and even dangerous. Their threat to human safety may result in their destruction. Should wildlife suffer for the pleasure you derive from feeding and petting them?"

The folks at the national parks also like to stress proper precautions when hiking in bear country because several people have been needlessly killed by bears in recent years. When you visit a park such as Yosemite or Denali you'll be bombarded with leaflets about bear behavior. Don't take them lightly because the information could save your life. Here are a few points to remember:

1. ***Odors attract bears*** Carry food in plastic bags or airtight containers and remove all garbage. Personal cleanliness is important, but don't use perfume, aftershave lotion or deodorant. A woman's menstrual period as well as human sexual activity may also attract bears.

2. ***Never surprise a bear*** Bears are secretive creatures by nature but will attack if they feel threatened. Wear bells, talk loudly, sing or whistle when hiking in bear territory. This gives the animals plenty of advance notice, and they will usually scamper off.

3. ***If you see a bear*** Slowly move away from the bear, keeping upwind so the bear realizes your presence from your scent. ***Don't run*** or make loud noises at this point because this could startle the bear.

4. ***Respect a mother bear and her cubs*** This combination is volatile if you disturb baby bears or simply come between the mother and cubs. A normally placid female will use violence to protect her offspring.

5. ***Climb a tree for safety*** If a bear charges at you, don't try to outrun it. The action provokes the chase instinct and bears can run faster than humans. Drop something sizeable (such as your pack) to divert the bear's attention and then climb the nearest tree.

6. ***If you are caught by a bear, play dead*** Lie on your stomach or side, with legs up against your chest and neck covered with clasped hands. The bear may pass by without inflicting any harm.

## *Security*

Unfortunately, the growing popularity of wilderness areas has increased the need for hikers to bolster measures for personal security. We read more and more accounts of backcountry robbers, rapists, and other undesirables. What can you do to protect yourself on the trail?

Some injuries (and deaths) are inflicted without criminal intent by careless hunters. The first step toward assuring that it's not you who gets shot, is to become familiar with the various hunting seasons where you live and avoid hiking the heavily hunted areas during those times. Ask local hunters or inquire at a sportman's shop. If you feel that an area is safe, stick to trail hiking. A hiker strolling along a nice path is less likely to be mistaken for an animal than a shapeless blob marauding through the underbrush.

Ordinarily I would advocate that hikers wear subtle earth-tone colors to minimize their visual impact on the environment, but during hunting season they should use the exact opposite philosophy when choosing their garb. Bright colors such as oranges and reds distinguish the hiker from the wildlife. And lastly, if you hear shots, consider turning back and rescheduling your outing for another time or place.

Women who hike in the wilderness are particularly vulnerable to premeditated crimes of violence, and a woman hiking alone is an easy target for a rapist. Her cries for help would probably not reach anyone. There's truth to the old adage about safety in numbers. Hike with one or more friends to discourage predators of the two-legged variety. Some women feel more comfortable if there are men in the group.

With all the hubbub over crime, backcountry and otherwise, would it be wise to carry a handgun? Only you can make that decision, based on the degree to which you would feel comfortable packing and using a pistol. Firearms are illegal in the national parks, but outside these boundaries their legality varies widely from state to state. Massachusetts and New York have strict gun control laws, while many Western states permit people to carry handguns as long as they are not concealed. You might consider carrying a whistle, which might be heard by someone nearby who could help you.

Some psychologists feel that people who carry guns derive a false sense of security and are therefore less cautious. Believing that they can pull a trigger to "solve" a problem should it arise, gun-toting backpackers may take risks and stumble into dangerous situations. In reality, they are inviting deeper trouble from a criminal who is more skilled and less hesitant to use a gun, such as the increasing number of armed men standing guard over illegally planted marijuana plots on public land. In this situation, you'd be far better off should you see marijuana growing, to get out immediately and report the location to the authorities.

The bottom line in protecting yourself from potentially dangerous individuals is awareness. Find out if there have been any negative incidents in the region. And don't hesitate to cancel your trip if you harbor any doubts about the people you see in the vicinity.

Just as you would desire that other people respect your own privacy and right to peacefulness while hiking, so too should you show the same respect to private landowners. If you would like to hike on someone's property, chances are the owner will not refuse a sincere request for permission. Let the landowner know exactly what your plans are: hiking, fishing, overnight camping, etc. Are there any areas which he would prefer you to stay away from such as planted fields or livestock areas? Asking questions will demonstrate that you intend to abide by his wishes and will insure a warm welcome in the future. Be sure to leave all gates in the position you found them, and if you must climb over a fence, do so near a fencepost, not at the middle where the support is weakest.

## *Weather Defenses*

The chances of encountering a bear or a mugger on the trail are remote, but weather is a factor the hiker must confront each and every time he goes into the wilderness. Under harsh situations an outdoorsperson can perish in a very short period of time. An understanding of some specific environmental conditions, and the problems your body will encounter on exposure to them, will ensure your safe return to civilization.

The higher you climb, the greater your chances of being struck by lightning during a storm. If you see or hear a storm brewing in the distance, plan ahead and seek a safe area before the storm strikes. Avoid standing on a mountain top or ridge, in an open field, or under a lone tree. A thick forest or deep cave offer safer refuge. You can also crouch between rocks in a boulder field. Should you be caught above timberline or in any other treeless location, place a non-conductive material such as a poncho or tarp on top of a small rock. Sit on this pad with hands clasped around your knees. Even if lightning hits you, the insulation may prevent the bolt from finding a path through your body.

The name hypothermia sounds like a rare jungle disease, but many outdoor experts consider it the greatest danger facing hikers today. Hypothermia is the abnormal lowering of internal body temperature. This can lead to mental confusion, physical collapse, and death. The phenomenon is dangerous because it can strike in the summertime when people are unprepared for cool weather. Although cold temperatures can induce the onset of hypothermia, the condition does not kill by freezing. Most cases develop in a temperature range of 30°-50°F.

The first stage of hypothermia begins when you lose heat faster than your body can produce it. You feel chilled. Involuntary tensing of muscles and shivering occurs in an effort to preserve heat in the vital organs. If the heat drain continues, the body's core temperature drops rapidly and hypothermia enters the second, and most critical, stage. Without being aware that it is happening, you lose judgment and reasoning ability. Next you are unable to control your hands. Soon you collapse into unconsciousness, and finally death.

The principle behind hypothermia prevention is simple: try to stay as warm and dry as possible by choosing protective clothing. Wool retains its insulating property even when wet. Rain gear should cover the head and neck as well as the torso to provide maximum protection against wind and rain. Dehydration can also contribute to hypothermia susceptibility, so you should drink plenty of water during a hike.

Be alert for the symptoms of hypothermia whenever your hiking party is exposed to wind, cold, and wetness. Paying attention to other members is important because the victim may be unaware of any problem. Look for shivering, slurred speech, lapse of memory, fumbling hands, difficulty walking, and drowsiness. If you spot any of these signs take immediate action. Move the victim to a sheltered area away from wind and rain. Take off all wet clothes and get the person into as many warm and dry clothes as possible. If you have a sleeping bag, wrap the victim inside. Give him or her warm (but not hot) drinks and high-energy foods such as candy. Although the hypothermia victim may feel drowsy and wish to sleep, it is extremely important to keep him or her awake. Sleeping could mean death.

Frostbite is a problem which can affect hikers in freezing temperatures, especially accompanied by wind. As blood supply to the extremities decreases, they become more susceptible to the cold. Watch for these symptoms in yourself and in others: whitened flesh, numbness, and blistering. Contrary to the advice of some people, you should not rub frostbitten skin with snow or anything else. This can damage the cells. Take off any items which tend to restrict circulation (boots, gloves, rings) and slowly rewarm the affected area with body heat from a hand, armpit or bare stomach of yourself or another companion. Keep the rest of the body dry and warm, and get indoors as soon as possible.

Hikers who climb to lofty heights may experience altitude sickness due to the lack of oxygen. The best prevention is to acclimatize your body by spending two or three days at the elevation before doing any hiking. When you do begin your trek, make the ascent gradual. After 9,000 feet you should probably not travel more than one thousand verticle feet per day. Severe headaches and nausea are signs that you may be suffering from altitude sickness. Take deep breaths, rest, and consume high-energy foods. Aspirin may ease the headaches. If the symptoms persist, you should descend to a lower elevation immediately. To remain at the same height in your weakened state is to ask for trouble.

When you deal with the outdoors on nature's terms, trouble can arise at any time. In the case of an emergency, the problem is immediate—a sprained ankle, a rabid animal. Often, however, problems creep up slowly and are not dealt with until it is too late. The more skilled you become at recognizing the warning signals, the better your chances of avoiding these hidden dangers. You should anticipate a bad situation before it affects you.

# Epilogue

One late winter day, a friend and I decided to go on a hike. Although the air was nippy, the sky was sunny and there was no snow on the trail to impede our footsteps. As I breathed in the cool air, the hiking soon became effortless and my feet seemed to float above the ground. Boy, did it feel good just to ***move***.

What a clever idea to leave the stale air indoors and go hiking in the middle of February, I told myself. I acted as if I had suddenly invented something new. Of course I hadn't. I was only feeling good about the nice experience that resulted from allowing spontaneity to take charge. Too often the danger in reading a how-to book such as this, is that it tends to focus our approach in such a structured and methodical way that we remove the spontaneity from the activity. Don't let this happen to you. Hiking is a can-do any time, anywhere activity. When you temper technique with instinct, hiking will remain pure and simple.

# APPENDIX 1: National Parks—United States

## Alabama

Horseshoe Bend National Military Park
Rt. 1, Box 103
Daviston, AL 36256

Russell Cave National Monument
Rt. 1, Box 175
Bridgeport, AL 35740

## Alaska

Denali National Park and Preserve
P.O. Box 9
McKinley Park, AK 99755

Katmai National Park and Preserve
P.O. Box 7
King Salmon, AK 99613

Glacier Bay National Monument
Bartlett Cove
Gustavus, AK 99826

Klondike Gold Rush National Historical Park
P.O. Box 517
Skagway, AK 99840

## Arizona

Canyon de Chelly National Monument
P.O. Box 588
Chinle, AZ 86503

Coronado National Memorial
Rt. 1, Box 126
Hereford, AZ 85615

Grand Canyon National Park
Grand Canyon, AZ 86023

Organ Pipe Cactus National Monument
Route 1, Box 100
Ajo, AZ 85321

Saguaro National Monument
Old Spanish Trail
Route 8, Box 695
Tucson, AZ 85730

Wupatki-Sunset Crater National Monuments
Tuba Star Route
Flagstaff, AZ 86001

Chiricahua National Monument
Dos Cabezas Star Route, Box 6500
Willcox, AZ 85643

Glen Canyon National Recreation Area
P.O. Box 1507
Page, AZ 86040-1507

Navajo National Monument
Tonalea, AZ 86044

Petrified Forest National Park, AZ 86028

Walnut Canyon National Monument
Rt. 1, Box 25
Flagstaff, AZ 86001

## Arkansas

Arkansas Post National Memorial
Rt. 1, Box 16
Gillett, AR 72055

Hot Springs National Park
P.O. Box 1860
Hot Springs National Park, AR 71901

Buffalo National River
P.O. Box 1173
Harrison, AR 72601

Pea Ridge National Military Park
Pea Ridge, AR 72751

## California

Channel Islands National Park
1901 Spinnaker Drive
Ventura, CA 93001

Death Valley National Monument
Death Valley, CA 92328

Golden Gate National Recreation Area
Fort Mason
San Francisco, CA 94123

Joshua Tree National Monument
74485 National Monument Drive
Twentynine Palms, CA 92277

Lassen Volcanic National Park
Mineral, CA 96063

Lava Beds National Monument
Tulelake, CA 96134

Muir Woods National Monument
Mill Valley, CA 94941

Pinnacles National Monument
Paicines, CA 94043

Point Reyes National Seashore
Point Reyes, CA 94956

Redwood National Park
1111 Second Street
Crescent City, CA 94431

Sequoia and Kings Canyon National Parks
Three Rivers, CA 93271

Whiskeytown-Shasta-Trinity National Recreation Area
P.O. Box 188
Whiskeytown, CA 96095-0188

Yosemite National Park
Yosemite National Park, CA 95389

## Colorado

Black Canyon of the Gunnison National Monument
P.O. Box 1648
Montrose, CO 81401

Colorado National Monument
Fruita, CO 81521

Curecanti National Recreation Area
P.O. Box 1040
Gunnison, CO 81230

Dinosaur National Monument
Dinosaur, CO 81610

Florissant Fossil Beds National Monument
P.O. Box 185
Florissant, CO 80816

Great Sand Dunes National Monument
Mosca, CO 81146

Hovenweep National Monument
McElmo Route
Cortez, CO 81321

Rocky Mountain National Park
Estes Park, CO 80517

## District of Columbia

National Park Service
National Capital Region
1100 Ohio Drive, SW
Washington, DC 20242

Rock Creek Park
5000 Glover Rd., NW
Washington, DC 20015-1098

## Florida

Big Cypress National Preserve
S.R. Box 110
Ochopee, FL 33943

Everglades National Park
P.O. Box 279
Homestead, FL 33030

Canaveral National Seashore
P.O. Box 2583
Titusville, FL 32780

Fort Caroline National Memorial
12713 Fort Caroline Rd.
Jacksonville, FL 32225

## Georgia

Chickamauga and Chattanooga
National Military Park
P.O. Box 2126
Fort Oglethorpe, GA 30742

Kennesaw Mountain National
Battlefield Park
P.O. Box 1167
Marietta, GA 30061

Cumberland Island National Seashore
P.O. Box 806
St. Marys, GA 31558

## Hawaii

Haleakala National Park
P.O. Box 537
Makawao, Maui, HI 96768

Puukohola Heiau National Historic
Site
P.O. Box 4963
Kawaihae, HI 96743

Hawaii Volcanoes National Park, HI
96718-0052

Pu'uhonua o Honaunau National
Historical Park
P.O. Box 128
Honaunau, Kona, HI 96726

## Idaho

Craters of the Moon National
Monument
Arco, ID 83213

## Indiana

Indiana Dunes National Lakeshore
1100 N. Mineral Springs Rd.
Porter, IN 46304

Lincoln Boyhood National Memorial
Lincoln City, IN 47552

## Iowa

Effigy Mounds National Monument
McGregor, IA 52157-0510

## Kentucky

Cumberland Gap National Historical Park
P.O. Box 840
Middlesboro, KY 40965

Mammoth Cave National Park
Mammoth Cave, KY 42259

## Maine

Acadia National Park
P.O. Box 177
Bar Harbor, ME 04609

## Maryland

Antietam National Battlefield Site
P.O. Box 158
Sharpsburg, MD 21782-0158

Catoctin Mountain Park
Thurmont, MD 21788

Assateague Island National Seashore
Rt. 2, Box 294
Berlin, MD 21811

Chesapeake and Ohio Canal National Historical Park
P.O. Box 4
Sharpsburg, MD 21782

## Massachusetts

Boston National Historical Park
Charlestown Navy Yard
Boston, MA 02129

Cape Cod National Seashore
South Wellfleet, MA 02663

## Michigan

Isle Royale National Park
Houghton, MI 49931

Sleeping Bear Dunes National Lakeshore
400 Main Street
Frankfort, MI 49635

Pictured Rocks National Lakeshore
Munising, MI 49862

## Minnesota

Grand Portage National Monument
P.O. Box 666
Grand Marais, MN 55604

Voyageurs National Park
P.O. Box 50
International Falls, MN 56649

Pipestone National Monument
Box 727
Pipestone, MN 56164

## Mississippi

Natchez Trace Parkway
Rural Route 1, NT-143
Tupelo, MS 38801

## Missouri

Ozark National Scenic Riverways
P.O. Box 490
Van Buren, MO 63965

Wilson's Creek National Battlefield
Postal Drawer C
Republic, MO 65738-0403

## Montana

Bighorn Canyon National Recreation Area
P.O. Box 458
Fort Smith, MT 59035

Glacier National Park
West Glacier, MT 59936

Custer Battlefield National Monument
P.O. Box 39
Crow Agency, MT 59022

## Nebraska

Agate Fossil Beds National Monument
P.O. Box 427
Gering, NE 69341-0427

Scotts Bluff National Monument
P.O. Box 427
Gering, NE 69341-0427

Homestead National Monument
R.F.D. #3, Box 47
Beatrice, NE 68310-9416

## Nevada

Lake Mead National Recreation Area
601 Nevada Highway
Boulder City, NV 89005

## New Jersey

Gateway National Recreation Area
Sandy Hook Unit
P.O. Box 437
Highlands, NJ 07732

Morristown National Historical Park
P.O. Box 1136R
Morristown, NJ 07960

## New Mexico

Bandelier National Monument
Los Alamos, NM 87544

Carlsbad Caverns National Park
3225 National Parks Highway
Carlsbad, NM 88220

White Sands National Monument
P.O. Box 458
Alamogordo, NM 88310

Capulin Mountain National Monument
Capulin, NM 88414

Chaco Culture National Historical Park
Star Route 4, Box 6500
Bloomfield, NM 87413

## New York

Fire Island National Seashore
120 Laurel Street
Patchogue, NY 11772

Gateway National Recreation Area
Breezy Point District
Building #1
Fort Tilden, NY 11695

Saratoga National Historical Park
R.D. #2, Box 33
Stillwater, NY 12170

## North Carolina

Blue Ridge Parkway
P.O. Box 9098-Oteen
Asheville, NC 28815

Cape Hatteras National Seashore
Rt. 1, Box 675
Manteo, NC 27954

Cape Lookout National Seashore
P.O. Box 690
Beaufort, NC 28516

Carl Sandburg Home National
Historic Site
P.O. Box 395
Flat Rock, NC 28731

Kings Mountain National Military
Park
P.O. Box 31
Kings Mountain, NC 28086

## North Dakota

Theodore Roosevelt National
Memorial Park
Medora, ND 58645

## Ohio

Cuyahoga Valley National Recreation
Area
P.O. Box 158
Peninsula, OH 44264

## Oklahoma

Chickasaw National Recreation Area
P.O. Box 201
Sulphur, OK 73086

## Oregon

Crater Lake National Park
Crater Lake, OR 97604-0007

John Day Fossil Beds National
Monument
420 W. Main
John Day, OR 97845

Oregon Caves National Monument
19000 Caves Highway
Caves Junction, OR 97523

## Pennsylvania

Allegheny Portage Railroad National Historic Site
P.O. Box 247
Cresson, PA 16630

Delaware Water Gap National Recreation Area
Bushkill, PA 18324

Fort Necessity National Battlefield & Friendship Hill National Historic Site
The National Pike
Farmington, PA 15437

Gettysburg National Military Park
Gettysburg, PA 17325

Hopewell Village National Historic Site
R.D. #1, Box 345
Elverson, PA 19520

Johnstown Flood National Memorial
c/o Allegheny Portage Railroad National Historic Site
P.O. Box 247
Cresson, PA 16630

Valley Forge National Historical Park
Valley Forge, PA 19481

## South Carolina

Ninety Six National Historic Site
P.O. Box 496
Ninety Six, SC 29666

## South Dakota

Badlands National Park
P.O. Box 6
Interior, SD 57750

Jewel Cave National Monument
Custer, SD 57730

Wind Cave National Park
Hot Springs, SD 57747

## Tennessee

Big South Fork National River & Recreation Area
Obed Wild and Scenic River
P.O. Drawer 630
Oneida, TN 37841

Great Smoky Mountains National Park
Gatlinburg, TN 37738

Shiloh National Military Park
Shiloh, TN 38376

Stones River National Battlefield
Rt. 10, Box 401
Old Nashville Highway
Murfreesboro, TN 37130

## Texas

Amistad National Recreation Area
U.S. Highway 90 West
Del Rio, TX 78840

Big Thicket National Preserve
P.O. Box 7408
Beaumont, TX 77706

Guadalupe Mountains National Park
3225 National Parks Highway
Carlsbad, NM 88220

Big Bend National Park
Rio Grande Wild and Scenic River
Big Bend National Park, TX 79834

Fort Davis National Historic Site
P.O. Box 1456
Fort Davis, TX 79734

## Utah

Arches National Park
Canyonlands National Park
Natural Bridges National Monument
National Park Service
P.O. Box 40
Monticello, UT 84535

Capitol Reef National Park
Torrey, UT 84775

Golden Spike National Historic Site
P.O. Box 394
Brigham City, UT 84302

Zion National Park
Springdale, UT 84767

Bryce Canyon National Park
Bryce Canyon, UT 84717

Cedar Breaks National Monument
P.O. Box 749
Cedar City, UT 84720

Timpanogos Cave National Monument
Rural Route Box 200
American Fork, UT 84003

## Virginia

Appomattox Court House National Historical Park
P.O. Box 218
Appomattox, VA 24522

Colonial National Historical Park
P.O. Box 210
Yorktown, VA 23690

George Washington Memorial Parkway
Turkey Run Park
McLean, VA 22101

Petersburg National Battlefield
P.O. Box 549
Petersburg, VA 23804

Shenandoah National Park
Luray, VA 22835

Booker T. Washington National Monument
Rt. 1, Box 195
Hardy, VA 24101

Fredericksburg National Military Park
P.O. Box 679
Fredericksburg, VA 22401

Manassas National Battlefield Park
P.O. Box 1830
Manassas, VA 22110

Prince William Forest Park
P.O. Box 208
Triangle, VA 22172

## Washington

Coulee Dam National Recreation Area
Box 37
Coulee Dam, WA 99116

North Cascades National Park
Sedro Woolley, WA 98284

San Juan Island National Historical Park
300 Cattle Point Road
Friday Harbor, WA 98250

Mount Rainier National Park
Tahoma Woods
Star Route
Ashford, WA 98304

Olympic National Park
600 East Park Avenue
Port Angeles, WA 98362

## West Virginia

Harpers Ferry National Historical Park
Harpers Ferry, WV 25425

## Wisconsin

Apostle Islands National Lakeshore
P.O. Box 729
Bayfield, WI 54814

St. Croix National Scenic Riverway
P.O. Box 708
St. Croix Falls, WI 54024

## Wyoming

Devils Tower National Monument
Devils Tower, WY 82714

Fossil Butte National Monument
P.O. Box 527
Kemmerer, WY 83101

Yellowstone National Park, WY 82190

Fort Laramie National Historic Site
Fort Laramie, WY 82212

Grand Teton National Park
Moose, WY 83012-0170

# *APPENDIX 2: National Parks—Canada*

## Alberta

Banff National Park
P.O. Box 900
Banff, Alberta
T0L 0C0

Jasper National Park
P.O. Box 10
Jasper, Alberta
T0E 1E0

Elk Island National Park
Site 4, R.R. 1
Fort Saskatchewan, Alberta
T8L 2N7

Waterton Lakes National Park
Waterton Park, Alberta
T0K 2M0

## British Columbia

Glacier National Park
P.O. Box 350
Revelstoke, British Columbia
V0E 2S0

Mount Revelstoke National Park
P.O. Box 350
Revelstoke, British Columbia
V0E 2S0

Yoho National Park
P.O. Box 99
Field, British Columbia
V0A 1G0

Kootenay National Park
P.O. Box 220
Radium Hot Springs, British Columbia
V0A 1M0

Pacific Rim National Park
Box 280
Ucluelet, British Columbia
V0R 3A0

## Manitoba

Riding Mountain National Park
Wasagaming, Manitoba
R0J 2H0

## New Brunswick

Fundy National Park
P.O. Box 40
Alma, New Brunswick
E0A 1B0

Kouchibouguac National Park
Kent County, New Brunswick
E0A 2A0

## Newfoundland

Gros Morne National Park
P.O. Box 130
Rocky Harbour, Newfoundland
A0K 4N0

Terra Nova National Park
Glovertown, Newfoundland
A0G 2L0

## Northwest Territories

Auyuittuq National Park
Pangnirtung, Northwest Territories
X0A 0R0

Nahanni National Park
Postal Bag 300
Fort Simpson, Northwest Territories
X0E 0N0

Wood Buffalo National Park
P.O. Box 750
Fort Smith, Northwest Territories
X0E 0P0

## Nova Scotia

Cape Breton Highlands National Park
Ingonish Beach, Nova Scotia
B0C 1L0

Kejimkujik National Park
P.O. Box 36, Maitland Bridge
Annapolis County, Nova Scotia
B0T 1N0

## Ontario

Georgian Bay Islands National Park
P.O. Box 28
Honey Harbour, Ontario
POE 1EO

Point Pelee National Park
R.R. No. 1
Leamington, Ontario
N8H 3V4

Pukaskwa National Park
P.O. Box 550
Marathon, Ontario
POT 2EO

St. Lawrence Islands National Park
P.O. Box 469, R.R. 3
Mallorytown Landing, Ontario
KOE 1RO

## Prince Edward Island

Prince Edward Island National Park
P.O. Box 487
Charlottetown, Prince Edward Island
C1A 7L1

## Québec

Forillon National Park
P.O. Box 1220
Gaspé, Québec
GOC 1RO

La Mauricie National Park
P.O. Box 758
Shawinigan Québec
G9N 6V9

### Saskatchewan

Prince Albert National Park
P.O. Box 100
Waskesiu Lake, Saskatchewan
SOJ 2YO

### Yukon Territory

Kluane National Park
Parks Canada
Haines Junction, Yukon Territory
YOB 1LO

# APPENDIX 3:

*State government agencies which distribute information about hiking trails*

State of Alabama
Department of Conservation
Administrative Division
Montgomery, AL 36130

Department of Natural Resources
Information Office
270 Washington Street SW, Rm 817
Atlanta, GA 30334

Illinois Department of Conservation
Life and Land Together
Lincoln Tower Plaza
524 South Second Street
Springfield, IL 62706

Iowa Conservation Commission
Wallace State Office Building
Des Moines, IA 50319

Department of Conservation
Bureau of Parks and Recreation
State House Station 22
Augusta, ME 04333

Department of Environmental Management
Division of Forests and Parks
100 Cambridge Street
Boston, MA 02202

Department of Conservation and Natural Resources
Division of State Parks
Capitol Complex #4162
Carson City, NV 89710

New York State
Department of Environmental Conservation
Albany, NY 12233-0001

State of Delaware
Department of Natural Resources and Environmental Control
P.O. 1401
Dover, DE 19901

Department of Land and Natural Resources
Division of Forestry and Wildlife
1151 Punchbowl Street
Honolulu, HI 96813

Division of Public Information
Indiana Department of Natural Resources
615 State Office Building
Indianapolis, IN 46204

Department of Parks
Capital Plaza Tower
Frankfort, KY 40601

Department of Natural Resources
Assistance and Information
Tawes State Office Building
Annapolis, MD 21401

Missouri Department of Natural Resources
P.O. Box 176
Jefferson City, MO 65101

State of New Hampshire
Office of Vacation Travel
P.O. Box 856
Concord, NH 03301

North Carolina
Department of Natural Resources and Community Development
P.O. Box 27687
Raleigh, NC 27611

North Dakota Game and Fish Department
2121 Lovett Avenue
Bismark, ND 58505

Department of Transportation
Parks and Recreation Division
525 Trade Street SE, Suite 301
Salem, OR 97310

South Dakota
Department of Game, Fish and Parks
Anderson Building
Pierre, SD 57501

Agency of Environmental Conservation
Department of Forests, Parks and Recreation
Montpelier, VT 05602

Ohio Department of Natural Resources
Divison of Geological Survey
Fountain Square
Columbus, OH 43224

Commonwealth of Pennsylvania
Department of Environmental Resources
P.O. Box 1467
Harrisburg, PA 17120

Department of Conservation
701 Broadway
Nashville, TN 37203

State of West Virginia
Department of Natural Resources
Charleston, WV 25305

# APPENDIX 4:

*Organizations which promote hiking and outdoor activities*

The Adirondack Mountain Club Inc.
172 Ridge Street
Glens Falls, NY 12801

Adirondack Trail Improvement Society, Inc.
St. Huberts, NY 12943

Alpine Club of Canada
P.O. Box 1026
Banff, Alberta
T0L 0C0

American Guides Association
Box 935
Woodland, CA 95695-0935

The American Hiking Society
1701 18th Street, NW
Washington, DC 20009

Appalachian Mountain Club
Five Joy Street
Boston, MA 02108

Appalachian Trail Conference
P.O. Box 236
Harpers Ferry, WV 25425

The Bruce Trail Association
P.O. Box 857
Hamilton, Ontario
L8N 3N9

Continental Divide Trail Society
P.O. Box 30002
Bethesda, MD 20814

The Florida Trail Association Inc.
P.O. Box 13708
Gainesville, FL 32604

Iowa Mountaineers
P.O. Box 163
Iowa City, IA 52244

Montana Outfitters and Guides Association
P.O. Box 631
Hot Springs, MT 59045

The Mountaineers
719 Pike Street
Seattle, WA 98101

National Campers and Hikers Association, Inc.
7172 Transit Rd.
Buffalo, NY 14221

National Trails Council
Box 493
Brookings, SD 57006

New England Trail Conference
c/o Forrest E. House
33 Knollwood Drive
East Longmeadow, MA 01028

Potomac Appalachian Trail Club
1718 N Street, NW
Washington, DC 20036

Randolph Mountain Club
Randolph, NH 03570

Skyline Hikers of the Canadian Rockies
P.O. Box 3514, Station B
Calgary, Alberta
T2M 4M2

Trail Riders of the Wilderness
American Forestry Association
1319 18th Street, NW
Washington, DC 20036

Sierra Club
530 Bush Street
San Francisco, CA 94108

# *APPENDIX 5:*

*Outdoor equipment suppliers*

L.L. Bean
3610 Main Street
Freeport, ME 04033

Black Ice
2310 Laurel Street
Napa, CA 94559

C.C. Filson Co.
205 Maritime Building
Seattle, WA 98104

Donner Mountain
2110 Fifth Street
Berkeley, CA 94710

Eastern Mountain Sports
2408 Vose Farm Road
Peterborough, NH 03458

Fjall Räven
Optimus, Inc.
Box 1950
Bridgeport, CT 06601

Helly-Hansen
Box C-31
Redmond, WA 98052

The North Face
1234 Fifth Street
Berkeley, CA 94710

REI Co-op
Box C-88126
Seattle, WA 98188

Sierra Designs
247 Fourth Street
Oakland, CA 94607

Summit Research
771 State Street
Schenectady, NY 12307

Because It's There
Box 4266
Pioneer Square Station
Seattle, WA 98104

Camp 7
802 S. Sherman
Longmont, CO 80501

Columbia Sportswear
Box 03239
Portland, OR 97203

Early Winters, Ltd.
110 Prefontaine Pl., S.
Seattle, WA 98104

Eddie Bauer
Fifth and Union
Dept. AY9
Seattle, WA 98124

Gerry
6260 Downing
Denver, CO 80216

JanSport/Downers
1466 Broadway, Suite 504
New York, NY 10017

Patagonia/GPIW
Box 150
Ventura, CA 93001

Royal Robbins
1314 Coldwell Avenue
Modesto, CA 95350

Sierra West
6 E. Yanonali Street
Santa Barbara, CA 93101

Terramar
Box 114
Pelham, NY 10803

Trailwise
Boss Manufacturing
221 W. First Street
Kewanee, IL 61443

Woolrich
Woolrich, PA 17779

Wilderness Experience
20675 Nordhoff
Chatworth, CA 91311

# APPENDIX 6:

*Bed & breakfast guide books*

**Bed and Breakfast American Style.** Norman T. Simpson. The Berkshire Traveller Press, Stockbridge, MA

**Bed and Breakfast U.S.A.: A Guide to Tourist Homes and Guest Houses.** Betty Rundback and Nancy Ackerman. E.P. Dutton, New York.

**Christopher's Bed and Breakfast Guide to the U.S. and Canada.** Bob and Ellen Christopher. Travel Discoveries, Milford, CT.

**Country Bed and Breakfast Places in Canada: A Guide to Warmth and Hospitality Along Canadian Highways and Byways.** John Thompson. The Berkshire Traveller Press, Stockbridge, MA.

**Country Inns and Back Roads North America.** Norman T. Simpson. The Berkshire Traveller Press, Stockbridge, MA.

**Country Inns, Lodges and Historic Hotels of the South.** Anthony Hitchcock and Jean Lindgren. Burt Franklin & Co., New York.

**The Great American Guest House Book.** John Thaxton. Burt Franklin & Co., New York.

# *READER COMMENTS*

Kids can be your harshest critics. I once got a letter from a *Boys' Life* reader who attempted unsuccessfully to follow my article directions for making a weather forecaster. He gave up in disgust, calling the project a "piece of junk." He also went on to say, "I'm never going to make anything from your stupid magazine again."

I hope no one will have such a negative reaction to this book, but whatever your feelings, I'd like to solicit your feedback. If you have any comments, criticisms or suggestions, drop me a note on this form. I've tried to make this a personal book between author and reader and your comments will let me know if I've succeeded. And if you have any questions, I'll try to answer them. Bye.

Mail to: David Lee Drotar
P.O. Box 515
Clifton Park, NY 12065

Book title: ***Hiking: Pure and Simple***

Dear Dave:

________________________________________

________________________________________

________________________________________

________________________________________

________________________________________

________________________________________

Name: ________________________

Address: ________________________

City, State, Zip: ________________________